£21.99

D1375640

Robotics, Mechatronics, and Artificial Intelligence

Experimental Circuit Blocks for Designers

Other Books of Interest by Newnes

ELECTRONIC CIRCUIT INVESTIGATOR SERIES

ROBERT J. DAVIS, *Digital and Computer Projects*

NEWTON C. BRAGA, *CMOS Projects and Experiments: Fun with the 4093 Integrated Circuit*

NEWTON C. BRAGA, *Electronic Projects from the Next Dimension: Paranormal Experiments for Hobbyists*

NEWTON C. BRAGA, *Pirate Radio and Video: Experimental Transmitter Projects*

RELATED TITLES

ANDREW SINGMIN, *Beginning Electronics through Projects*

ANDREW SINGMIN, *Practical Audio Amplifier Circuit Projects*

ANDREW SINGMIN, *Beginning Digital Electronics through Projects*

Robotics, Mechatronics, and Artificial Intelligence

Experimental Circuit Blocks for Designers

Newton C. Braga

Newnes

Boston Oxford Auckland Johannesburg Melbourne New Delhi

Newnes is an imprint of Butterworth–Heinemann.

Copyright © 2002 by Butterworth–Heinemann

 A member of the Reed Elsevier group

All rights reserved.

Recognizing the importance of preserving what has been written, Butterworth–Heinemann prints its books on acid-free paper whenever possible.

ISBN 0-7506-7389-3

British Library Cataloguing-in-Publication Data
A catalogue record for this book is available from the British Library.

The publisher offers special discounts on bulk orders of this book.
For information, please contact:

Manager of Special Sales
Butterworth–Heinemann
225 Wildwood Avenue
Woburn, MA 01801-2041
Tel: 781-904-2500
Fax: 781-904-2620

For information on all Newnes publications available, contact our World Wide Web home page at:
http://www.newnespress.com

10 9 8 7 6 5 4 3 2 1

Printed in the United States of America

Contents

Acknowledgments xvii

About the Author xix

Chapter 1   Fundamentals of Robotics and Mechatronics...................... 1

1.1   Purpose... 1
1.2   History 1
1.3   The Structure of Robotics and Mechatronics Projects...................... 3
    1.3.1   Control 4
    1.3.2   Actuators 4
1.4   Using Basic Blocks in Projects 9
1.5   Suggested Projects 9
1.6   Additional Information... 14
1.7   Review Questions... 15

Chapter 2   Motion Controls.. 17

2.1   Purpose... 17
2.2   Theory 17
    2.2.1   Direction.. 17
    2.2.2   Speed 17
2.3   Characteristics 18
    2.3.1   Voltage... 18
    2.3.2   Current 18
    2.3.3   Power 18
    2.3.4   Speed 19
2.4   Basic Blocks... 19
    Block 1:   Simple DC Control 22
    Block 2:   Reversing the Direction 22
    Block 3:   Two-Way Control....................................... 22
    Block 4:   Controlling Two Motors with One Switch (I) 23
    Block 5:   Controlling Two Motors with One Switch (II)................ 23
    Block 6:   Half-Wave Rectifier 24
    Block 7:   Full-Wave Control 24
    Block 8:   Two-Speed Block Using Diodes........................... 25
    Block 9:   Multi-speed Control Using Diodes........................ 25
    Block 10:   Two-Speed Voltage Change from the Source 26
    Block 11:   Power Booster... 26
    Block 12:   Multi-step Power Booster............................... 27
    Block 13:   Adding Inertia... 27

Block 14:   Series–Parallel Switch 28
Block 15:   Motors in Series and Parallel 29
Block 16:   LED Direction Indicator 29
Block 17:   Current Indicator 30
Block 18:   Adding Sound (I) 30
Block 19:   Adding Sound (II) 31
Block 20:   Two-Wire × Two-Motor Control 32
2.5   Suggested Projects 32
2.6   Additional Information 32
2.7   Review Questions 33
Chapter 3   Controlling Motors, Relays, and Solenoids with Transistors 35
3.1   Purpose 35
3.2   Theory 35
    3.2.1   The Relay 35
    3.2.2   Basic Blocks Using Relays 36
    Block 21:   Turning a Load On or Off 37
    Block 22:   Current Reversion 37
    Block 23:   Series–Parallel Switching (I) 37
    Block 24:   Series–Parallel Switching (II) 38
    Block 25:   Timed Relay 39
    Block 26:   Short-Action Relay 39
    Block 27:   Flasher 40
    Block 28:   Passive Buzzer 40
    Block 29:   Locked Relay 41
3.3   The Transistor as a Switch 41
    3.3.1   Bipolar Transistors 41
    3.3.2   Field Effect Transistor or FETs 43
3.4   Practical Blocks 44
    Block 30:   Switching with an NPN Transistor 44
    Block 31:   Switching with a PNP Transistor 45
    Block 32:   Using an NPN Darlington Pair 45
    Block 33:   Using a PNP Darlington Pair 46
    Block 34:   Using a Darlington NPN Transistor 46
    Block 35:   Using a Darlington PNP Transistor 47
    Block 36:   Complementary Driver (I) 47
    Block 37:   Complementary Driver (II) 47
    Block 38:   Power FET Driver 48
3.5   More Blocks Using Relays and Transistors 48
    Block 39:   Delayed Relay 49
    Block 40:   Timed Relay 49
    Block 41:   Long-Interval Delayed Relay 49
    Block 42:   Long-Interval Timed Relay 50
    Block 43:   Delayed Relay Using Power FET 50
    Block 44:   Holding Circuit 51
    3.5.1   Working with the Blocks 52
    Block 45:   Reversing a Motor for a Few Seconds 52

Block 46:   Reversing and Stopping a Motor 53
Block 47:   Reversing a Motor and Reducing Its Speed 54
3.6   Suggested Projects 54
3.7   Additional Information. 54
   3.7.1   Information about Relays 54
   3.7.2   Information about Transistors 56
3.8   Review Questions. 57

Chapter 4   H-Bridges 59
4.1   Purpose. 59
4.2   Theory 59
   4.2.1   Half Bridge. 59
   4.2.2   H-Bridge or Full Bridge 60
4.3   Practical Circuits 62
Block 48:   Simple Half Bridge Using Symmetric Supply 63
Block 49:   Half Bridge Using Darlington Transistors 64
Block 50:   Full Bridge Using Complementary Bipolar Transistors 65
Block 51:   Full H-Bridge with Feedback 66
Block 52:   Full Bridge Using NPN Transistors 66
Block 53:   Full Bridge Using NPN Darlington Transistors 67
Block 54:   Full Bridge with Logic Control. 68
Block 55:   Full Bridge with Logic Control and External Enable 69
Block 56:   Power MOSFET H-Bridge 69
Block 57:   H-Bridge Using Power MOSFETs with Enable 70
Block 58:   Combined Bipolar + MOSFET H-Bridge. 71
Block 59:   Combined Darlington + Power MOSFET H-Bridge 72
Block 60:   R-S Flip-Flop H-Bridge 72
Block 61:   Complete H-Bridge. 72
4.4   Integrated H-Bridges 74
Block 62:   H-Bridge Using the LMD18200. 74
Block 63:   H-Bridge Using the LM18201 76
4.5   Additional Information. 76
4.6   Special Recommendations 78
   4.6.1   Decoupling Capacitors 78
   4.6.2   Transistor Protection 79
   4.6.3   Current Sensing 80
   4.6.4   Fuses. 80
4.7   Suggested Projects 80
4.8   Review Questions. 81

Chapter 5   Linear and PWM Power Controls 83
5.1   Purpose. 83
5.2   Theory 83
   5.2.1   Two Types of Controls. 84
   5.2.2   Linear Controls. 84
   5.2.3   Pulse Width Modulation. 84
   5.2.4   Two Forms of PWM Controls 86
5.3   Basic Blocks. 87

5.3.1   Linear Controls. 87
   Block 64:   Electronic Rheostat. 88
   Block 65:   Linear Control Using a Darlington Transistor 88
   Block 66:   Linear Power Control Using a Zener Diode 89
5.3.2   Constant Current Sources. 90
   Block 67:   Constant Current Source Using Transistor 90
   Block 68:   Constant Current Source Using the LM350T (3 A) 91
   Block 69:   Variable Constant Current Source Using the LM338 (5 A). 92
5.3.3   PWM Blocks 92
   Block 70:   PWM Basic Control Using the 4093 CMOS IC. 92
   Block 71:   PWM Control Using the 4001/4011 CMOS IC 92
   Block 72:   Medium/High-Power PWM Control Using the 555 IC 94
   Block 73:   Medium-Power PWM Using the 555 IC and a PNP Transistor 94
   Block 74:   Anti-phase PWM Power Control Using the 555 IC 94
   Block 75:   Power Anti-phase PWM Control Using the 555 IC 96
   Block 76:   PWM Control Using the LM350 96
5.4   Additional Information. 97
5.5   Suggested Projects 98
5.6   Review Questions. 98

Chapter 6   Power Control Using Thyristors 99
6.1   Purpose. 99
6.2   Theory 99
   6.2.1   Silicon Controlled Rectifiers 99
   6.2.2   Triacs 101
   6.2.3   Other Devices of the Thyristor Family 102
6.3   Basic Blocks Using SCRs 105
   Block 77:   Turning a Load On and Off with an SCR. 105
   Block 78:   Delayed Turn-On Switch with SCR 106
   Block 79:   Touch Switch Using an SCR 106
   Block 80:   Triggering an SCR with Positive Pulses. 107
   Block 81:   Triggering SCRs with Negative Pulses. 107
   Block 82:   Crowbar Protection. 107
   Block 83:   Overcurrent Protection 109
   Block 84:   R-S Flip-Flop Using an SCR 109
   6.3.1   SCRs in AC Circuits 110
   Block 85:   Simple AC Switch. 110
   Block 86:   Full-Wave AC Switch (I) 110
   Block 87:   Full-Wave AC Switch (II) 111
   Block 88:   Dimmer and Speed Control 111
   Block 89:   Full-Wave Dimmer. 113
   Block 90:   Dimmer Using UJT and SCR 113
   6.3.2   Blocks Using Triacs. 114
   Block 91:   High-Power AC Switch 114
   Block 92:   High-Power Dimmer. 114
6.4   Additional Information. 115
   6.4.1   Care when Using Inductive Loads 115

6.4.2   Characteristics of Some Common SCRs and Triacs 115
6.5   Suggested Projects 116
6.6   Review Questions. 116

Chapter 7   Solenoids, Servomotors, and Shape Memory Alloys 119
7.1   Purpose. 119
7.2   Theory 119
    7.2.1   Shape Memory Alloys 119
    7.2.2   The Solenoid 121
    7.2.3   The Servomotor 122
7.3   Practical Blocks 124
    7.3.1   Blocks for Electronic Muscles (SMAs) 124
        Block 93:   Simple Drive Circuit for SMAs 124
        Block 94:   Rheostat for SMA. 125
        Block 95:   Constant Current Source. 125
        Block 96:   NPN Transistor Drive for SMA 126
        Block 97:   Using a Bipolar PNP transistor. 126
        Block 98:   Driving SMA from Power MOSFETs 126
    7.3.2   Blocks for Solenoids 127
        Block 99:   Turning a Solenoid On and Off 127
        Block 100:   Determining the Force of a Solenoid 127
        Block 101:   Intelligent Control for Two Solenoids 128
        Block 102:   Intelligent Solenoid Control Using Darlington Transistors 129
        Block 103:   Intelligent Control Using CMOS Logic 130
        Block 104:   Intelligent Circuit Using CMOS IC and Power CMOS 130
        Block 105:   Current Sensing (I) 130
        Block 106:   Current Sensing (II) 131
    7.3.3   Blocks for Servos 133
        Block 107:   Servo Control Using DC Motors 133
        Block 108:   Control for R/C Standard Servos 134
7.4   Additional Information. 134
7.5   Suggested Projects 134
7.6   Review Questions. 135

Chapter 8   Stepper Motors. 137
8.1   Purpose. 137
8.2   Theory 137
    8.2.1   How It Works. 138
8.3   How to Use Stepper Motors. 140
    8.3.1   Voltage and Current 140
    8.3.2   Sequence. 141
    8.3.3   Step Angle 141
    8.3.4   Pulse Rate. 141
    8.3.5   Torque 142
    8.3.6   Braking Effect 142
8.4   Blocks Using Stepper Motors 142
        Block 109:   Standard Block Using Bipolar NPN Transistors 142
        Block 110:   Standard Block Using Darlington NPN Transistors. 142

Block 111: Driving a Stepper Motor with PNP Transistors 143
Block 112: Using PNP Darlington Transistors 144
Block 113: Driving a Stepper Motor with Power MOSFETs 145
Block 114: Step Generator Using the 555 IC 145
Block 115: Step Generator Using the 4093 IC 146
Block 116: Step-by-Step Generator. 147
Block 117: Control for Two-Phase Stepper Motor 147
8.4.1   Using ICs 147
Block 118: Driving a Stepper Motor with the ULN2002 and ULN2003 148
Block 119: Driving Stepper Motors with the MC1413/MC1416 148
Block 120: Complete Stepper Motor Control with the SAA1027. 148
Block 121: Complete Stepper Motor Control Using the UCN4202 150
Block 122: LED Monitor for Stepper Motor Operation 151
8.5   Additional Information. 152
8.6   Suggested Projects 152
8.7   Review Questions. 153

Chapter 9   On-Off Sensors. 155
9.1   Purpose. 155
9.2   Theory 155
9.2.1   Debouncing 157
9.2.2   Switches as Sensors 158
9.2.3   Reed Switches 158
9.2.4   Home-Made Sensors 159
9.2.5   Programmed or Sequential Mechanical Sensors 159
9.3   Basic Blocks Using Sensors. 159
Block 123: Turning a Load On 160
Block 124: Turning a Load Off. 160
Block 125: Contact Conditioner with a Capacitor. 160
Block 126: Low-Current Turn-On Sensor with Contact Conditioner. 161
Block 127: Low-Current Turn-Off Sensor With Contact Conditioner 161
Block 128: Contact Conditioner Using the 555 IC 162
Block 129: Contact Conditioner Using the 4093 CMOS IC (I) 163
Block 130: Contact Conditioner Using the 4093 CMOS IC (II). 163
Block 131: TTL Contact Conditioner Using the 7400 IC 163
Block 132: Contact Conditioner for Two Sensors—Bistable 164
Block 133: Contact Conditioner for SPDT Sensor 164
Block 134: Contact Conditioner for Two Sensors. 165
9.3.1   Notes on Compatible Blocks 165
9.3.2   Controlling Motors and Loads 165
Block 135: Direction Control Using a Mechanical Sensor 165
Block 136: Controlling Two Loads. 166
Block 137: High-Low Sensor 166
Block 138: High-Power Motor Control. 166
Block 139: Timed Sensor Using a 555 IC. 167
Block 140: High-Current Control for Mechanical Sensors. 168
Block 141: Multi-voltage Control for Mechanical Sensors. 168

Block 142: Priority Switch 169

Block 143: Tachometric Sensor 170

Block 144: Missing Pulse Detector.................................. 170

9.4   Suggested Projects 171

9.5   Additional Information... 171

9.6   Review Questions.. 173

Chapter 10   Resistive Sensors 175

10.1   Purpose... 175

    10.1.1   Theory 175

    10.1.2   The LDR or CdS Cell................................... 175

    10.1.3   Negative Temperature Coefficient Resistors 176

    10.1.4   Pressure Sensors....................................... 177

    10.1.5   Potentiometers as Position Sensors 177

    10.1.6   Touch Sensors 177

10.2   How to Use Resistive Sensors 178

10.3   Practical Blocks 179

    Block 145: Basic Resistive Sensor Circuit (I)...................... 179

    Block 146: Basic Resistive Sensor Circuit (II) 179

    Block 147: Basic Block Using a PNP Transistor (I)................. 179

    Block 148: Basic Block Using PNP Transistor (II).................. 180

    Block 149: Differential Sensor 180

    Block 150: Snap Action for Resistive Sensors (I)................... 181

    Block 151: Snap Action for Resistive Sensors (II) 182

    Block 152: Increasing Sensitivity 182

    Block 153: Light-Activated Circuit Using an SCR.................. 183

    Block 154: Dark-Activated Circuit Using an SCR 184

    Block 155: Priority Circuit Using Resistive Sensors................. 184

    Block 156: Opto-Isolator Using LDR 185

    Block 157: Opto-Isolator with Logic Input (TTL and CMOS)............ 185

    Block 158: Current Sensor Using an NTC 186

    Block 159: Thermal Crowbar 187

    Block 160: Light/Temperature-Controlled Oscillator................. 187

    Block 161: Light-Sensitive/Temperature-Dependent Oscillator........... 187

    Block 162: Light/Temperature-Dependent Oscillator.................. 188

    Block 163: Light/Temperature-Triggered Monostable 188

    Block 164: Fast Monostable Sensor 189

    Block 165: Light/Temperature-Dependent Oscillator.................. 190

10.4   Additional Information... 191

10.5   Suggested Projects 191

10.6   Review Questions.. 191

Chapter 11   Operational Amplifiers and Comparators 193

11.1   Purpose... 193

11.2   Theory 193

    11.2.1   Operational Amplifiers and Comparators 193

    11.2.2   The Window Comparator............................... 195

    11.2.3   How to Use Operational Amplifiers and Comparators 196

11.2.4   Choosing an Opamp 197
11.2.5   Power Supplies 198
11.3   Practical Blocks 199
   Block 166: Voltage Follower 199
   Block 167: Amplifier with Gain 199
   Block 168: Driving an NPN transistor 200
   Block 169: Driving a PNP Transistor 201
   Block 170: Basic Voltage Comparator 201
   Block 171: Negative Voltage Comparator 202
   Block 172: Using Resistive Sensors (I) 202
   Block 173: Using Resistive Sensors (II) 203
   Block 174: Resistive Sensors Driving PNP Transistor 203
   Block 175: Differential Sensor 204
   Block 176: Touch Sensor/Pressure Sensor 205
   Block 177: Delayed Turn-On Relay 205
   Block 178: Driving TTL Blocks 206
   Block 179: Power Comparator 206
   Block 180: Low-Frequency Squarewave Oscillator 206
   Block 181: Double Comparator 207
   Block 182: Step Comparator 208
   Block 183: Window Comparator (I) 209
   Block 184: Window Comparator (II) 209
11.4   Additional Information 210
11.5   Suggested Projects 211
11.6   Review Questions 211

Chapter 12   Remote Controls and Remote Sensing 213
12.1   Purpose 213
12.2   Theory 213
   12.2.1   Wires 213
   12.2.2   Light 213
12.3   Infrared 214
   12.3.1   Sound 214
   12.3.2   Radio 215
   12.3.3   Electromagnetic Interference 215
12.4   Choice and Use of a Remote Control 216
12.5   Blocks 216
   Block 185: Remote Control Using Wires 216
   Block 186: Multi-wire Control 217
   Block 187: Sequential Control Using Wires 218
   Block 188: Matrix Control 219
   Block 189: Tone Generator 220
   Block 190: AC Switch (I) 221
   Block 191: AC Switch (II) 221
   Block 192: PLL Tone Decoder 222
   Block 193: Multi-tone Transmitter 223
   Block 194: Multichannel Tone Decoder 223

Block 195: Remote Control Using the AC Power Line, Transmitter 224
Block 196: Remote Control Using the AC Power Line, Receiver 224
Block 197: Flashlight Remote Control 225
Block 198: Infrared Transmitter 226
Block 199: Infrared Receiver 227
Block 200: Sound/Ultrasonic Transmitter. 227
Block 201: Low-Impedance Sound/Ultrasonic Receiver 228
Block 202: Audible Receiver for Electret Microphone. 228
Block 203: 30 to 100 MHz Transmitter Block 229
Block 204: Using a Small FM Radio as Remote Control Receiver 230
Block 205: 30 to 100 MHz Super-Regenerative Receiver 230
Block 206: Coder Using the MC145026. 230
Block 207: Decoder Using the MC145026. 230
Additional Information 232
Suggested Projects. 232
12.6   Review Questions. 234

Chapter 13   Logic Blocks **235**
13.1   Purpose. 235
13.2   Theory 235
        13.2.1   Gates, Inverters, and Buffers 235
        13.2.2   Monostable/Bistable. 237
        13.2.3   Counters and Decoders. 238
        13.2.4   Voltage-Controlled Oscillators 238
        13.2.5   Memory 238
13.3   How to Use Logic Blocks 238
13.4   Blocks. 239
        Block 208: NOR Gate Using a Transistor. 239
        Block 209: NAND Gate Using a Transistor 239
        Block 210: Logic Inverter Using a Transistor. 240
        Block 211: Basic Monostable Using the 555 IC 240
        Block 212: Sequential Monostable 241
        Block 213: Multi-timing Using the 555 IC 241
        Block 214: Bistable Block. 241
        Block 215: Complete Bistable Using the 4013 IC 242
        Block 216: Frequency Divider 243
        Block 217: Frequency Divider Using the 4020 IC 244
        Block 218: Divide-by-10 Counter Using the 4017 IC. 244
        Block 219: Divide-by-n Using the 4017 IC. 245
        Block 220: Programming with Diodes 246
        Block 221: Digital-to-Analog Converter (DAC). 246
        Block 222: Sequential Programming Using a Diode Matrix. 247
        Block 223: Voltage-Controlled Oscillator. 247
        Block 224: Logic Switch. 249
        Block 225: Diode Matrix. 249
        Block 226: R/2R Network—DAC. 249
        Block 227: Binary Coded Decimal Counter/Decoder. 250

Block 228: Complete DAC 251
Block 229: 64-Bit RAM 251
13.5   Additional Information. 253
13.6   Suggested Projects 254
13.7   Review Questions. 255

Chapter 14   Intelligence and the Computer 257
14.1   Purpose. 257
14.2   Theory 257
   14.2.1   What is Intelligence? 257
   14.2.2   Intelligence by Hardware and Software. 258
   14.2.3   Electronic Neurons and Neural Networks 258
   14.2.4   Fuzzy Logic and Intelligent Software 259
   14.2.5   Microcontrollers and Microprocessors. 259
   14.2.6   Personal Computers 260
14.3   Using the Hardware 260
14.4   Blocks. 260
   Block 230: Learning Circuit Using Power MOSFET 260
   Block 231: Learning Circuit Using an IC 261
   Block 232: Sample-and-Hold Circuit 261
   Block 233: High-Power Learning Circuit 262
   Block 234: Light-Conditioned Circuit. 262
   Block 235: Thermal Memory 263
   Block 236: Teachable Window Comparator 264
   Block 237: Electronic Neuron (I) 265
   Block 238: Electronic Neuron (II). 265
   Block 239: Learning Circuit Using the 4017 IC 266
   Block 240: Integrate-and-Fire Neuron Using Operational Amplifiers 266
   14.4.1   Connecting a Computer 267
   Block 241: Simplest Parallel Interface 268
   Block 242: Parallel Interface (II). 268
   Block 243: Isolated Interface (I) 268
   Block 244: Isolated Interface (II) 269
   Block 245: Interface for AC-Powered Loads 269
   Block 246: Data Acquisition Interface Using Comparators 270
   Block 247: Data Acquisition Interface Using the ADC0808 270
14.5   Additional Information. 271
14.6   Other Microcontrollers. 272
14.7   Suggested Projects 272
14.8   Review Questions. 272

Chapter 15   Other Blocks: Light and Sound Effects. 273
15.1   Purpose. 273
15.2   Theory 273
   15.2.1   Light Effects. 273
   15.2.2   Sound Effects 274
   15.2.3   Other Blocks. 274
   15.2.4   Self-Defense 275

15.2.5   Power Supplies. 275
15.2.6   Battery Chargers. 275
15.3   Blocks. 275
Block 248: Low-Voltage LED Flasher 276
Block 249: Block 317–LED Lamp Flasher. 276
Block 250: Fluorescent Lamp Inverter 277
Block 251: Fluorescent Lamp Flasher. 278
Block 252: LED Sequencer. 278
Block 253: Driving Incandescent Lamps 279
Block 254: Bar/Dot Display Driver. 280
Block 255: Variable-Sound Siren 281
Block 256: Dual-Tone Siren 281
Block 257: Sound and Light 281
Block 258: High-Voltage Defense 282
Block 259: Solenoid Gun 283
Block 260: Laser Gun 284
Block 261: General-Purpose Power Supply 284
Block 262: Battery Charger. 285
15.4   Additional Information. 286
15.4.1   Musical Modules 286
15.4.2   Other Circuits. 287
15.5   Suggested Projects 287
15.6   Review Questions. 288

Chapter 16   Working Safely. 289
16.1   The Importance of Safety. 289
16.2   Experimental and Industrial Robots. 289
16.3   Safety Rules 291

Answers to Review Questions 293

Acknowledgments

I would like to thank all the people who helped make this book possible: my agent, Jeff Eckert, who has helped in more ways than I can mention; Isabel Pereira da Silva, who worked hard to produce the drawings; Renato Palott, who was helped prepare the files; the editors at Newnes; and my wife, who has been supportive of my efforts.

About the Author

Mr. Braga was born in São Paulo, Brazil, in 1946. He received his professional training at São Paulo University (USP). His activities in electronics began when he was 13 years old, at which time he began to write articles for Brazilian magazines. At age 18, he had his own column in the Brazilian edition of *Popular Electronics,* where he introduced the concept of "electronics for youngsters."

In 1976, he became technical director of the most important electronics magazine in South America, *Revista Saber Eletrônica* (published in Brazil, Argentina, and Mexico). He also has been the director of other magazines published by the same company, including *Eletrônica Total.* During this time, Mr. Braga has published more than 60 books about electronics, computers, and electricity, and thousands of articles and electronic projects in magazines all over the world (U.S.A., France, Spain, Portugal, Argentina, Mexico, et al.). More than 2,000,000 copies of his books have been sold throughout Latin America and Europe.

The author also teaches electronics and physics and is engaged in educational projects in his home country of Brazil. These projects include the introduction of electronics in secondary schools and professional training of workers who need enhanced knowledge in the field of electronics. The author now lives in Guarulhos (near São Paulo) and is married, with a 10-year-old son.

1

Fundamentals of Robotics and Mechatronics

1.1 Purpose

The purpose of this first part of our book is to explain the basics of robotics, mechatronics, and artificial intelligence and to describe how to make practical projects using basic building blocks. With an understanding of how robots, automatic systems, mechatronics projects, and intelligent machines are built, the reader can assemble a customized system, creating a functional project that can be used for practical purposes, experimentation, education, public exhibitions, or just as an exercise of the imagination.

This text explores only the practical aspects of electronic circuits used in robotics and mechatronics, leaving deep theory to more advanced books. Herein the reader will not find much in the way of mathematical theory or discussions of the physics of how a chosen project will operate after assembly. Instead, we focus on building blocks from which the reader can develop a practical, and sometimes empirical, project. The correct values for components and mechanical parts can be derived from experience.

1.2 History

Robotics, mechatronics, and *artificial intelligence* are words that have become very popular these days, although their definitions sometimes are controversial. The idea of machines that can serve all our needs, freeing us from difficult, repetitive, and dangerous tasks, and the concept of electronic servants that can work for us unceasingly without protests, strikes, or simple fatigue, have been deeply explored by science fiction and movies.

Hugo Gernsback, Isaac Asimov, and many other science fiction authors are very popular among those who have explored the themes of robots, automatic machines, and artificial intelligence. In particular, the works of Hugo Gernsback, who used electronics in many of his narratives, are recommended to the reader who is interested in this subject. You can visit his web site at **http://www.hugogernsback.com.**

To trace the origin of robotics, we must travel in time, going back as far as ancient Greece, where the first movable statues were built. In the first century B.C., Hero de Alexandria conducted experiments with mechanical birds. From some time later, around 270 B.C., we have accounts of Cresibus, a Greek engineer who made organs and water clocks with movable figures.

Still later, in A.D. 770, the Swiss clock maker Pierre Jacquet-Droz created three mechanical dolls that could play music on an organ, draw simple pictures, and write. Another name that cannot be omitted from the history of robotics is Nicola Tesla, who built a radio-controlled submarine.

The word *robot* was coined in 1921 by the Czechoslovakian novelist Karel Kapec in his book, *R.U.R.—Rassum's Universal Robots*. In this book, he described mechanical servants that could do all the things that a man could. *Robot* is just the Czech word for *worker*. From that time until today, the word has become a term used to indicate all mechanical beings that can perform some tasks normally done by humans. More recently, many authors have explored the idea of mechanical beings, including the famous R2-D2 and C-3PO seen in movies of the Star Wars series.

The popular image of a robot is normally associated with creatures having forms or behaviors that resemble those of humans. The human-like robots in most cases have arms, legs, and a head that can think, and sometimes they even display some "emotions." For the scientists and engineers who work with projects in robotics, the practical image of a working robot is otherwise.

The basic idea of a practical robot is to replace humans in repetitive, dangerous, or tiring tasks. Because the most repetitive and tiring task found in the real word today is the manufacture of objects on a production line, the first idea of practical robotics was associated with industry. *Industrial robots* were the first real "beings" of this category to enter our modern word as operating machines, making things for us. Cars, electronic devices, and home appliances today are manufactured by automatic machines or "robots" that replace the common worker in a production line, much to our advantage.

The interesting point to consider is that a practical robot on an industrial production line has an appearance that, in most cases, doesn't resemble a human being in any way. These devices have no legs and no head, but many arms that are designed in appropriate forms and sizes for specific tasks. There is no practical necessity to make them resemble human appendages.

Of course, the idea of human-like robots that work in your home as servants, or that perform other practical tasks, has not disappeared. From this idea—the association of electronics with mechanics to produce machines that can perform a wide range of tasks automatically—has come a new science called *mechatronics*.

Many definitions of mechatronics have been offered. The basic idea is the use of a *synergistic integration* of mechanics, electronics, and computer technology to produce enhanced products or systems. Mechatronics thus is a subset of *cybernetics*.

Since mechatronics and robotics can be placed in parallel blocks when studying their applications and circuits, they share many common points. This is why, when analyzing the curriculum of many courses in mechatronics, we find that they cover the operation and construction of robots. Likewise, when analyzing the curriculum of robotics courses, we find that they deal with the construction of devices that integrate mechanics and electronics!

Another subject related to these studies is *artificial intelligence*. Modern robotics and mechatronics designs often include some degree of intelligence. The question of what criteria should be used to define *intelligence* in this case, and to determine what level exists within a machine, is a problem that lends itself to many approaches.

For example, Alan Turing, a british mathematician, conceived an interesting test to determine if a machine is or isn't intelligent. He proposed the use of a person to test the machine. Man and machine are connected via some kind of a modem such that the person doesn't know who is at the other end of the line. Through this connection, they can exchange messages, thereby conducting a conversation. If the machine can maintain a conversation well enough that the person can't be sure if it is a human or machine, the machine can be considered intelligent.

Today, engineers, researchers, experimenters, and students, when thinking about robotics, mechatronics, or artificial intelligence, have many common issues to consider. Projects in this field have some structural similarities, since they all use electronics and mechanics. (Because the aim of this book is to provide electronic building blocks for these projects, we are not including hydraulics in this field. But this is a real possibility that should be considered for more advanced concepts.) This means that the same basic blocks can be used in any project in any of these fields. The way the blocks are organized is the only the factor that determines what the machine will do. In the next section, we will see how the robotics, mechatronics, and artificial intelligence blocks can be used in practical projects.

1.3   The Structure of Robotics and Mechatronics Projects

Our point of departure is the idea that robots and mechatronics devices are machines that use electronics and mechanics in combination, and that they are created to perform tasks normally performed by people. From this basic assumption, we can separate the required functions into blocks. The general structure is shown in Figure 1.1.

The number and selection of blocks used in a specific project are determined by the end result envisioned by the designer. A fixed arm, for instance, or an au-

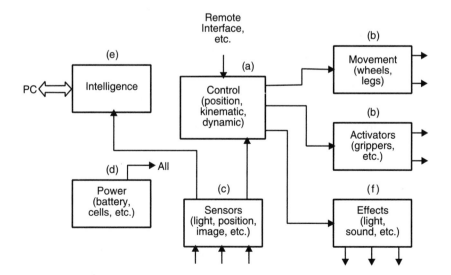

Figure 1.1   Basic organization of a project in mechatronics or robotics.

tomatic elevator doesn't need wheels or legs. A head with electronic eyes programmed to "see" and detect things doesn't need arms. The common blocks
used in all the projects are as described below.

1.3.1 Control

This is the "brain" of any project in a robotic or mechatronic system. All the
electric parts of a robot or any other project are controlled by electronic circuits.
The basic controls for robots and mechatronic projects are as follows.

(a) Position control. Arms with grips or other object manipulators must have
precise controls to place them in the correct position. The movement of a head
with eyes is controlled by a one-axis control block.

(b) Kinematic control. Any project with moving parts needs this kind of control. The speed of any moving part must be precisely determined and controlled
by the circuits. One of the most important of the controls in this group governs
the speed of the motor that moves a robot.

(c) Dynamic control. Many parts of a robot or mechatronic project incorporate forces that must be controlled during operation. When the grip of a robot
manipulates an object, its necessary to use some kind of control that determines
what degree of force is necessary to grasp the object firmly without breaking it.
A real challenge for the project builder is to build a robot grip that can take an
egg from one basket and put it in another without breaking it. Such tasks require
precise dynamic control.

(d) Adaptive control. Adaptive control is needed when a function of a robot
or mechatronic device must change during the execution of a process. An example is the need for increased grip force when pressing against a spring. More
force is required as the spring is compressed. Another example of adaptive control is the application of more power to a motor to maintain a constant speed, as
when a robot moves from level ground to an incline or when it must move a
heavy object.

(e) External control. External controls are used when a human operator is
used to command all the tasks of a robot. The human is the "brain," and it uses
such sensors (senses) as vision in the control of the robot's operations.

To transfer commands for a robot or mechatronic device, the operator can use
a variety of "interfaces." The basic options are links that employ radio, infrared
(IR), wires, and even voice commands. Modern projects can include voice recognition circuits to receive orders directly from an operator. A computer can
also be used to interface the control unit and the robot or mechatronic device.

At this point, is important to consider the degree of intelligence that may exist in a
robot. Complex controls can give an observer the false impression that the robot is
"intelligent." However, a control block that employs many functions is not an intelligent block. Intelligence can be added if the robot must make decisions based in input
from its own sensors or from an operator who employs a specific data input block.

1.3.2 Actuators

Robots and mechatronic machines must have some way to manipulate objects
or perform some kind of action in the external world. As outlined below, many
types of actuators are found in practical projects.

Movement. Robots can move from one place to other using legs, wheels, or tracks. The legs can be moved using motors, solenoids, or shape memory alloys* (SMAs).

Manipulation. Robots and mechatronic devices don't have hands. They use grippers to manipulate objects, and these grippers are controlled by electronic circuits. The movement of the grippers can be driven with solenoids, motors, or SMAs. Figure 1.2 shows some types of grippers.

Manipulation can also be effected using equipment that is specially designed for a specific task, as we see in many industrial robots. In many applications, mechanically coupled parts can be adapted to fit the size and shape of whatever object must be manipulated. This is suggested by Figure 1.3.

Sensors. Robots and mechatronic devices detect what happens in the real world using sensors. The sensors are very important, since they can transmit information about the position of a robot or an arm, the size and shape of an object being manipulated, the presence of obstacles (if it is a moving robot), and much other information, including the object recognition by size and shape, as found in some "intelligent" projects.

A TV camera can be linked to an intelligent circuit, thus allowing an automatic arm to select pieces that have a specific size and shape from among many mixed pieces, as shown in Figure 1.4.

The basic sensors found in projects of robotics and mechatronics are as follows:

- *Light:* Light-dependent resistors (LDRs, e.g., CdS cells or photoresistors), photodiodes, photo cells, and phototransistors
- *Pressure:* conductive foam, electromechanical sensors, semiconductor sensors

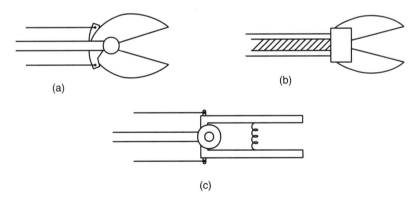

(a)

(b)

(c)

Figure 1.2   Various gripper types.

---

* Shape memory alloys are extraordinary materials that, via a process known as *martensitic transformation,* return to a predetermined shape when heated. When an SMA is cold (i.e., below transformation temperature), it can be deformed easily into a new shape. When the material is heated above its transformation temperature, it undergoes a crystal structure change that causes it to return to its original shape. If an SMA meets with resistance to this transformation, it can generate very strong forces. The most common shape memory material, called Nitinol, is a 50:50 alloy of nickel and titanium.

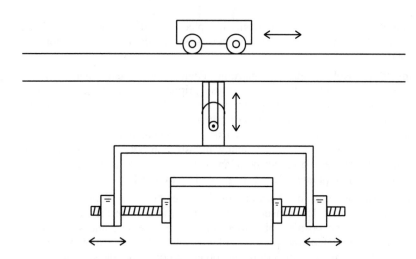

Figure 1.3 Manipulators adapted for the size and shape of the object.

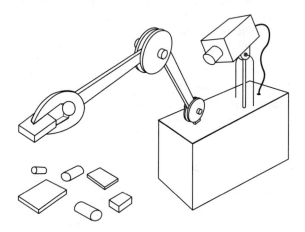

Figure 1.4 TV camera used an in image sensor.

- *Temperature:* NTC, PTC, diodes, or transistors[*]
- *Image:* Charge-coupled devices (CCDs), photodiodes, or phototransistor matrices
- *Position:* potentiometers, sonar, radar, infrared (IR) sensors
- *Contact:* microswitches, pendulums
- *Proximity:* capacitive, inductive, or IR

Power. Any project involving electronics and moving parts needs some kind of electric power supply. If the project is a moving robot, it is ideal if you can put

---

[*] NTC = negative temperature coefficient, PTC = positive temperature coefficient. With an NTC device, resistance goes down as temperature increases. A PTC device works in the opposite manner.

the power supply inside it. Batteries and cells can be used in this way. The size and the type of battery depends on the power needed by the robot, how long it must operate without recharging, and the tasks it must do.

If the device is fixed (e.g., a robotic arm) or relatively fixed (e.g., an elevator), electric power can be supplied by the ac power line. Since the electronic circuits and electromechanical devices used in the project usually use dc power, circuits that provide rectification, filtering, and voltage stabilization must be included if an ac power source is used. Alternative power sources, including solar cells, generators coupled to internal combustion engines, fuel cells, and many others, can be employed in some projects.

Intelligence.   This is in important block that is used in many mechatronics and robotics projects. Intelligence often can be considered as an independent block.

The intelligence blocks process information that is picked up by sensors or received from other external sources (e.g., a computer or the human operator) and make decisions about the tasks that the system must perform. The intelligence block can be as simple as a basic neural comparator that, for example, senses ambient light and determines the robot's distance from the light source or a wall. It can also encompass much more complex configurations, involving high-level decisions.

Two forms of artificial intelligence are suitable for applications in robotics and mechatronics:

Software intelligence.   Software intelligence is provided by a computer, microprocessor, or microcontroller in which any intelligent software runs. Hardware links provide the data the processor needs to make decisions and communicate with the control block.

The decisions are programmed in a basic structure and in some cases can be changed according to the incoming data. In such a case, the program can "learn" with experience, which is considered to be a basic characteristic of intelligent systems.

Programs that simulate neural networks are the preferred by students, researchers, and developers in the artificial intelligence field. Another important tool for the design of intelligent systems is *fuzzy logic.*

Software intelligence can be located inside the own robot or mechatronic machine when microprocessors and microcontrollers are used. The BASIC Stamp® chip* provides a simple way to add some degree of intelligence to a machine. It can be programmed to take some decisions from the inputs of sensors of external control.

The experimenter can find many programs that simulate neural networks and fuzzy logic. Many can be used to add intelligence to computers, automatons, and other machines.

Hardware intelligence.   Another way to add intelligence to a machine is by using circuits that can learn. The basic idea is to imitate the way living beings process the information they receive via their senses, i.e., using the nervous system. The analogy is shown in Figure 1.5.

---

* BASIC Stamps are small computers that run PBASIC programs. They are available from Parallax, Inc. (parallaxinc.com).

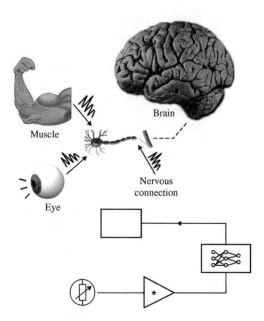

Figure 1.5   Adding intelligence via a neural network.

The living being receives the information from the external world by sensors coupled to nervous cells. The cells send the information to a brain, formed by a network of cells, where it is processed. The result is sent back to effectors (muscles, for instance) by another network of nervous cells.

Using artificial neurons, the designer can wire the elements together, forming neural networks that imitates the operation of the living being's information processing system, adding some degree of intelligence to a robot.

Of course, a single electronic neuron can't produce what we can define as an intelligent behavior. It is the behavior of an amoeba. Even using thousands or millions of neurons, the intelligence you will add to your project will not exceed that of a worm.

A neural network can learn and make decisions, making the machine somewhat intelligent. Neural networks in the form of integrated circuits can be bought from component dealers.[*] Starting with ICs of this type, designers can create complex neuron networks with such capabilities as learning, decision making, and more.

Effects.   Depending on the application, the robot or the mechatronic device must have some special means of getting people's attention, or which may even simulate human behavior. In practical projects, we can consider including blocks that produce sounds, speak words or answer questions, produce light effects, or even provide a self-defense mechanism.

Basically, the effects that can be used in these projects involving robotics and mechatronics are as follows:

---

[*] An example is the NNP® chip, from Accurate Automation Corp., www.accurate-automation.com.

- Sounds: sirens, voice synthesis, heartbeat, breathing, etc. (The heartbeat and breathing are used in some dolls to add a bit of realism.)
- Light effects: flashers, blinkers, stroboscopic signaling, etc.
- Defense: high voltage.

1.4   Using Basic Blocks in Projects

As we said in the introduction, the basic idea of this book is describe basic electronic building blocks for each of the many function that can be used in a project involving robots and mechatronic devices. With an understanding of how each basic block functions, and how it can be coupled to other blocks, the reader can comprehend the entire operation of an existing mechanism or create an original machine.

As mentioned, the number of blocks and their functions depend on the task the reader intends the machine to perform. When using the basic blocks, the designer must have in mind some important points.

Connectivity. The blocks must be designed so as to allow them to be interconnected without problems. This means that the signal sourced by the output of one block must be compatible with the input needs of other blocks. This book takes this fact into consideration, and most of the circuits can "talk" one with each other, as they have matched characteristics.

Compatibility. The designer of robotics and mechatronics projects must be attentive to the fact that the block he intends to use for a particular task must really have the ability to do it. A block that can control only 1 A across a load cannot be used to control a 2 A load. The designer must be sure that the block used to control a gripper can draw the necessary current to drive it.

Power drain. If the project is powered from a battery or cell, it is important to determine whether the battery can really provide enough current for all the blocks. Motors, actuators, and other transducers are the responsible for most of the energy drain in mechatronics and robotics projects.

The main problem occurs when all the transducers and actuators are activated at the same time. The high current drain can cause a large enough voltage drop to affect the operation of the entire system. It is a good idea is to use independent power sources for high current loads (such as motors, grippers, and actuators) and the electronic circuits.

If the circuit is powered from the ac power line, the same precautions are valid: use a large enough power supply to support all the devices that must be powered. It is also a good idea to use one power supply for motors, actuators, and other heavy duty loads, and another to power the electronic circuits.

1.5   Suggested Projects

At this point, the reader should be ready to conceive a first project in mechatronics or robotics and commit the idea to paper. The functions your machine will have, and what you intend it to do, can be laid out using basic blocks such as the ones we have already discussed. Using the basic blocks shown by Figure 1.6 you can design a functioning machine. Some projects are suggested in the following paragraphs.

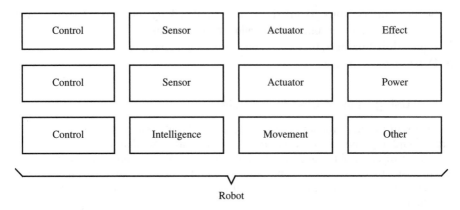

Figure 1.6 Blocks are ganged to form a complete project.

Automatic Arm

This simple mechatronic project consists of a robot arm or mechanism that can pick up objects from one place and put them in a basket or another location, as suggested by Figure 1.7. The basic configuration in blocks is shown in the same figure and includes the following functions:

- *Control.*   Obtaining information from a keypad or a joystick.
- *Actuator.*   Moving the arm and the gripper. An alternative for the gripper in a simplified project is the use of an electromagnet, if the objects to be moved can be attracted by it.
- *Sensor.*   Optionally, we can send information to the operator such as whether the basket is empty or full, whether there are obstacles, and if the object falls before it is placed in the basket.
- *Power supply.*
- *Effects.*   Light or sound effects can be added; a light can flash or a counter can be ratcheted up when the object falls into the basket.

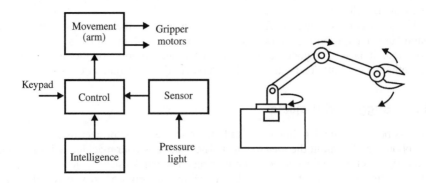

Figure 1.7 Automatic arm and block diagram.

Walking Robot

This is certainly the most attractive among all the projects the designer of robotics and mechatronics projects can adopt. Figure 1.8 show the basic blocks to form this kind of robot. Its functions are as follows:

- *Control.* The signals for this block can be sent by the operator (via radio control, IR link, or a cable), or they can be generated by sensors engaged by the moving system (motor, wheels, tracks, or legs). If the robot has other movable parts (e.g., arms or a head), they can be controlled by this block.
- *Brain.* Some intelligence can be added with the aid of sensors. The robot can make some decisions independently from the operator if we use a logic block for this task.
- *Sensors.* The sensors can inform the operator about the presence of obstacles and instruct the brain to make required decisions.
- *Actuators.* This robot can be equipped with a grip or other actuator to move objects. An "artificial hand" can be adapted from toys such as the one shown in Figure 1.9.
- *Effects.* Light and sound effects can be added to this robot. A solid state audio recorder can be used to make it speak some words or send some kind of warning message while walking around. The effects are very important if the robot is to be demonstrated at fairs, public exhibitions, or school contests.
- *Power.* This kind of robot is powered from batteries.

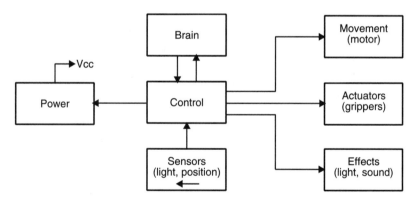

Figure 1.8   Walking robot, basic building blocks.

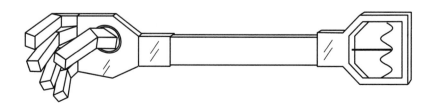

Figure 1.9   Plastic robot arm; gripper can be used in many projects.

Robot—Follow the Line

An interesting and classic project is the "follow-the-line" robot. It is equipped with LDRs or phototransistors as sensors in a differential "intelligent" configuration as shown by Figure 1.10.

The light sensors are excited by the light reflected by a white line drawn in the ground such that the control circuit of the robot can direct it to follow the line, as suggested by Figure 1.11.

In the basic version, the line forms a closed loop, but the designer can add some other sensory blocks to make it seek out the line and then follow it when found.

The basic blocks for this robot are:

- *Sensor.*   You can use LDRs or phototransistors.
- *Control.*   This block is operated directly from signals from the sensor or the intelligence block.
- *Intelligence.*   This block is formed by a neural comparator that senses and compares the light picked up by the sensors.
- *Power.*   A battery is the best solution but, for experimental purposes in a lightweight protoype, common cells are suitable.
- *Effects.*   Light and sound effects can be added if the reader deems them to be important.

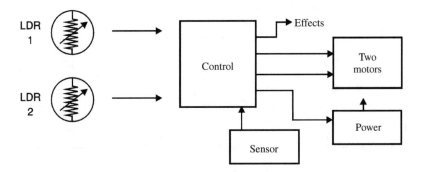

Figure 1.10   "Follow-the-line" robot configuration.

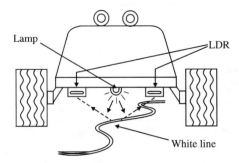

Figure 1.11   Line sensors "see" the line, controlling the robot.

Automatic Elevator

The automatic elevator is another classic project in mechatronics and automation. As Figure 1.12 shows, it is formed by creating a model of a building with an elevator. The number of floors can be as many as ten in a typical project.

The aim of the project is automatic operation using sensors at every floor. A keyboard is used to call the elevator to a specific floor, and an intelligence block can be added if we want it to make special decisions. This block can make the elevator "decide" which floor to service first if two calls are made at the same time. The decision depends on the floor at which it is located at that moment. The automatic elevator can be designed using the blocks shown in Figure 1.13.

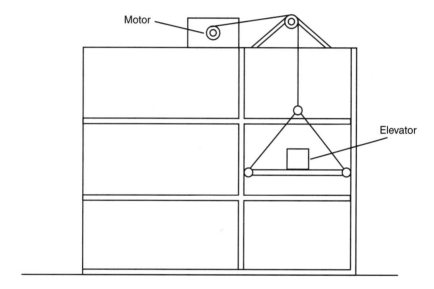

Figure 1.12   The automatic elevator.

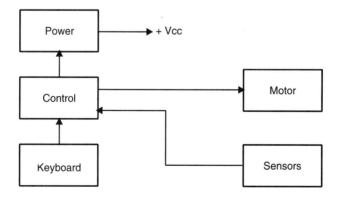

Figure 1.13   Building blocks of the automatic elevator.

Create Your Own Project

The following are some suggestions for projects you can create, preparing the blocks and their interconnections in advance:

1. A scale model production line that can manipulate some products
2. An automatic arm that can recognize and pick up ferromagnetic objects (iron, steel, etc.) and drop them into a basket
3. A robot that can follow a light beam from a flashlight
4. A robot that can find a sound source (your whistle, for example)
5. A scale model automatic car wash
6. A human-like robot that can speak some words to you and follow you around

Many contests for robot builders involve such tasks as a playing football, conducting a battle (one robot must destroy the other), movement of objects, etc. On the Internet, the reader will find many sites that deal with robotics and contests. Type "robot" or "robot contests" in a search engine such as Altavista, Yahoo, etc., and you may be surprised at the amount information that comes back.

1.6   Additional Information

This book is intended to give basic concepts, in the form of electronic blocks, for the design of mechatronic and robotic machines. Of course, the reader who intends to build these devices needs to have some basic knowledge of electronics and also must own some basic tools. The following paragraphs outline the prerequisites.

Basic Foundation in Electronics

The reader need not be an engineer or have a Ph.D. in electronics to design projects in robotics or mechatronics. It is important to note that most of the projects in this field are produced by high school and college students, and also by hobbyists who have never attended a technical school.

To build the electronic components of these projects, the reader must be somewhat skilled with tools and electronic components and must know something about the following:

- How components and basic circuits work
- How to read a schematic diagram
- How to make a printed circuit board from a diagram
- How to use basic instruments (e.g., a multimeter) to test the circuits if they do not function as expected, or to adjust their parameters for a specific task
- Where to find the components

Many existing books about basic electronics can benefit a reader who does not presently have the required skills and knowledge. For this reader, we recommend starting with very simple projects and progressing later to more complex ones.

Tools

The basic tools the reader needs to build the electronic circuits shown in our building blocks are these:

- Soldering iron (15 to 40 W). Suitable models include a dual-wattage pencil iron (Radio Shack 64-2060) or a 15 W pencil (Radio Shack 64-2052) and a 25 W pencil (Radio Shack 64-2072).
- Cutting pliers or diagonals (often called *dykes*) in sizes from 4 to 6 in.
- Chain-nose or needle-nosed pliers with narrow tips in sizes up to 5 in.
- Screwdrivers sized between 2 and 8 in. (Radio Shack 64-1823, a seven-piece set).
- Crimping tools: stripper and cutter for 10 to 22 wire ga. (Radio Shack 64-2129).
- Precision tool set with small screwdrivers of hex, common, and Phillips types (Radio Shack 64-1948).
- Soldering and desoldering accessories such as a desoldering bulb (Radio Shack 64-2086) and a soldering iron holder/cleaner (Radio Shack 64-2078).
- Extra grippers to hold the components and boards (Radio Shack 64-2094) or a project holder (Radio Shack 64-2093).
- Miniature hand drill (Radio Shack 64-1779, a five-piece hand drill set).

Instruments

- A simple analog multimeter (1 kΩ/V or more). If you are more experienced with the use of more advanced instruments, you may also have a digital multimeter and an oscilloscope on your bench.
- A PC can also be considered as a development tool for designers. Its interfaces for data acquisition can be used to analyze the signals from the circuits and to control them.

1.7   Review Questions

1. What is the difference between robotics and mechatronics?
2. Explain the origin of the word *robot.*
3. What is SMA, and how is it used in robotics and mechatronics?
4. What kind of electronic sensors can be used as "eyes" in a robot?
5. What is the difference between software and hardware intelligence?
6. What is the Turin test to determine whether a machine is intelligent?
7. What is dynamic control in a robot?
8. Give examples of appliances in your home that are based on mechatronics.
9. What is a neural network?
10. What is a gripper?

2

Motion Controls

2.1 Purpose

This purpose of this chapter is to show how switches (including relays and sensors) can be used to control dc motors and how capacitors, diodes, and other passive components can be used to add performance to dc motors.

2.2 Theory

The simplest way to add motion to robots, mechatronic devices, and other automation equipment is to use a dc motor. They are cheap, small, efficient, and can be found in a wide range of sizes, shapes, and power ratings.

A conventional dc motor is formed by an arrangement of coils and magnets that creates motion for electric power. Figure 2.1 shows the typical construction of a small dc motor.

We can use these motors to move any mechanism directly or, by adding gears or tracks, to reduce speed or increase power. The basic dc motor properties that must be considered when using them in robotics or mechatronics projects are as discussed below.

2.2.1 Direction

When powered from a dc power supply (battery or other source), the direction of the shaft rotation depends on the direction of the current flowing through the motor. by reversing the current, we can reverse the direction of movement of any device driven by a dc motor.

2.2.2 Speed

The speed of a dc motor, expressed in rotations per minute (rpm), depends on the current and the load. We must consider two situations when using a dc motor. In the

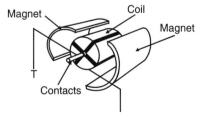

Figure 2.1 Small dc motor using permanent magnets.

first case, the motor operates without a load (or with a constant load). The speed will increase to a maximum that depends on the applied voltage.

In the second case, the motor operates with a variable load (i.e., the motor must power some kind of mechanism with forces that depend on the moment or the task). In this case, the speed depends on the load: higher required power = lower speed.

2.3   Characteristics

When looking for a dc motor for a particular application, the designer must determine its characteristics. The following are important considerations.

2.3.1   Voltage

Small dc motors can be obtained with voltage ratings in a range from 1.5 to 48 V. The specified voltage indicates the *nominal* voltage, which is the applied voltage that makes it run normally (i.e., producing maximum power and consuming nominal current). In practice, the nominal voltage is important, because it indicates the maximum recommended voltage that can be applied to the motor.

2.3.2   Current

The current flow through a motor, when powered with the nominal voltage, depends on the load. The current increases with an increase in load. It is important to avoid letting a motor run with excessive loads that can stall it. In the stalled condition, the motor becomes a short circuit for the current, and all applied power is converted into heat. The motor can quickly burn up in this condition. Common dc motors have operating currents in the range of 50 mA to more than 2 A.

2.3.3   Power

The power is given by the product of voltage × current. In projects involving robots and mechatronics, it is normal to rate the amount of force that a motor can generate in terms of its *torque* (twisting power). As shown by Figure 2.2, the torque is the force released by the shaft, and it depends not only on the motor's electrical and mechanical characteristics but also on the shaft diameter. This specification is important, as the force that a mechanism powered from a dc motor can provide depends not only on the motor itself but also on the mechanism coupled to it. Therefore, if gearboxes are added, as shown in Figure 2.3, the speed can be decreased, and the power will be increased by the same factor.

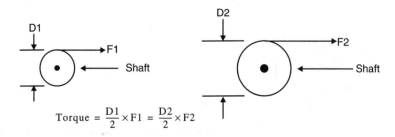

$$\text{Torque} = \frac{D1}{2} \times F1 = \frac{D2}{2} \times F2$$

Figure 2.2   Torque is constant for a motor.

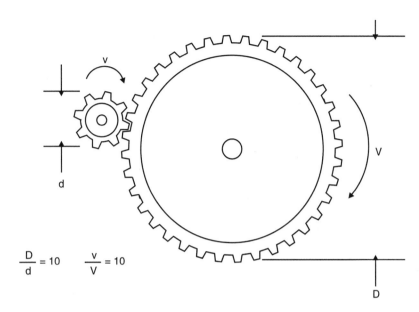

$$\frac{D}{d} = 10 \qquad \frac{v}{V} = 10$$

Figure 2.3   Reducing speed and increasing force with gears.

For instance, if a gearbox employs a reduction gear that is ten times the diameter of the motor drive gear, the speed is reduced ten times, but the power is also increased by a factor of ten.

In robotics and mechatronics, the use of gearboxes of all sizes and reduction ratios are commonly used to match the characteristics of a selected motor to a desired task. Therefore, when using a motor, is important to know the torque rating because, starting from that figure and knowing the reduction rate of the mechanism to be moved, we can easily find the final power that can be generated by a system.

2.3.4 Speed

It is normal to specify the speed of a dc motor in open or no-load conditions. The speed can be in the range of 500 and 10,000 rpm according to the type, size, and other characteristics of a common dc motor. Remember that this speed will be reduced by a considerable factor when the motor is operating under loaded conditions.

2.4   Basic Blocks

Starting from the idea that a dc motor can be controlled by the voltage applied to it and that

- direction depends on the current flow
- speed depends on the voltage applied or the current flow

we can devise applications that used the basic components described below.

SPST switch.   An SPST (single pole, single throw) switch is a device that can turn the current on and off in a circuit, as shown by Figure 2.4. The same function can be performed by a relay (as we will see in the following paragraphs).

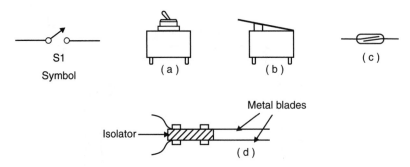

Figure 2.4   SPST switch configurations.

The SPST can be found in many forms, as shown by the same figure. Cross-switches (b) can be activated by small forces and coupled to sensors. Magnetic switches or reed switches (c) are activated by small magnets, and in (d) we show a sensor made by two contacts or wires that act as an SPST and can be used in many blocks described in this book.

DPDT switch.   A common DPDT (double pole, double throw) switch is shown in Figure 2.5. By combining the poles provided by this device, we can change the way a load is connected to a circuit. The main application for this switch is in reversing the current flow to select parallel/serial modes of operation of a load. A relay can also have contacts that perform the functions of a DPDT switch.

Single pole, multiple throw switches.   This kind of device allows the operator to select one of many circuits to be powered from a source, as shown in Figure 2.6. This type of switch can be found in several forms, including the *rotary* switch.

Figure 2.5   DPDT switch.

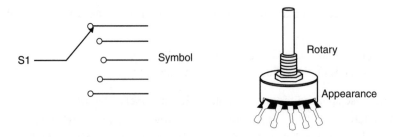

Figure 2.6   Single pole, multiple throw switch.

Semiconductor diodes. In a semiconductor diode, a current flow can be established only in one direction. When *forward biased,* a diode presents a low resistance, and current can flow. When *reversed biased,* the diode presents a high resistance, and no current can flow. This is illustrated in Figure 2.7.

Diodes are specified by their maximum operation current and voltage. They also can be specified by a manufacturer's part number such as the 1N4002 (used in many applications and rated for 50 V × 1 A).

These diodes can be used to convert alternating currents to direct currents, in logic functions and in protection circuits. Observe that diodes are *polarized* devices. Their correct position in a circuit must be observed. In general, diodes have their poles (anode and cathode) identified as shown in the figure: a ring is printed on the anode side.

Capacitors. Capacitors are devices that are used to store electrical energy. They are formed by metal plates with a *dielectric* (isolator) between them.

The amount of energy that a capacitor can store is given by its capacitance rating as measured in Farads (F). In practice, we use capacitors with ratings much lower than 1 F. To represent these capacitances, we use of submultiples of the farad. Common measurement units are the microfarad (μF), representing 0.000,001 F; the nanofarad (nF), representing 0.000,000,001 F; and the picofarad (pF), representing 0.000,000,000,001 F. Figure 2.8 shows the symbol and the appearance for several types of capacitors.

In many cases the capacitors are indicated by the type of dielectric they use. A ceramic capacitor, for instance, uses a piece of ceramic as the dielectric.

The other important specification for a capacitor is the maximum voltage that can applied across its plates. This is the *working voltage,* sometimes abbreviated as WVDC (working voltage dc).

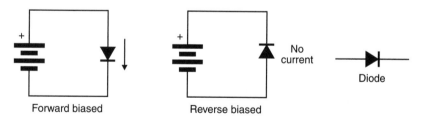

Figure 2.7   Biasing a diode.

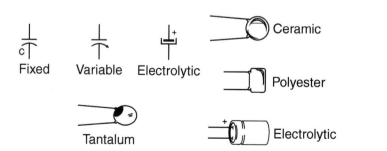

Figure 2.8   Appearance and symbols for various capacitor types.

Block 1   Simple DC Control

The simplest way to control a dc motor (or other load) is by adding an SPST switch in series as shown in Figure 2.9. When the switch is open, no current flows, and the motor is off. With the switch closed, the current can flow across the motor, and it is on. Note that the switch can be wired in series with the circuit in either the positive or the negative line.

Block 2   Reversing the Direction

The direction in which the shaft of a motor moves depends on the direction of the current. We can reverse the current direction with a DPDT switch as shown by Figure 2.10. The power supply represented in this case is a battery, but you can use any dc voltage source.

Block 3   Two-Way Control

The basic circuit shown in Figure 2.11 is very interesting for robotics and mechatronics applications. With this block, we can control a motor from two different points or sensors. Any of the switches (relays or sensors) can turn the motor on or off independently.

The circuit is the same one used to control lights from two switches in the wall. Special two-way switches must be used in this application, or sensors with two contacts, as suggested by Figure 2.12. In a mobile robot, one sensor can be place in the

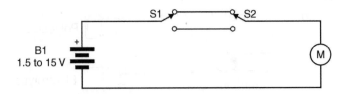

Figure 2.9   Block 1: simple dc control.

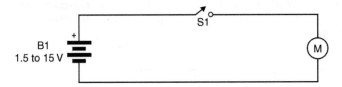

Figure 2.10   Block 2: reverse circuit.

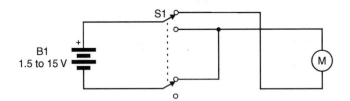

Figure 2.11   Block 3: two-way control.

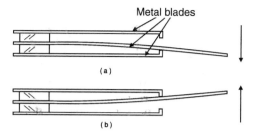

Figure 2.12   Switch made with metal blades.

front and another in the back. These can be used to detect obstacles and feed information to a decision block.

Block 4   Controlling Two Motors with One Switch (I)

This block can be used to control two motors from one switch or sensor. With the switch in position A, motor M1 is on, and with the switch in position B, motor M2 is powered.

Combining this block with Blocks 1 and 2, we have a complete control for two motors—reversing directions, turning them on and off, and selecting which one is activated. Block 4 uses a single pole, double throw switch as shown in Figure 2.13.

Block 5   Controlling Two Motors with One Switch (II)

The difference between this circuit and Block 4 is that we can use only a pair of wires to control two motors instead three as in the other configuration. The circuit shown in Figure 2.14 uses diodes, selecting the motors to be powered according to the direction of the current flow. When switch S1 is in position 1, D1 is forward biased, and motor M1 is powered on. When S1 is placed in position 2, D2 is forward

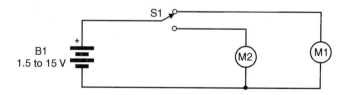

Figure 2.13   Block 4: controlling two motors with one switch (I).

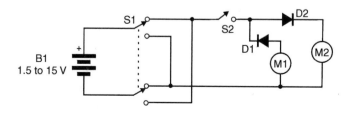

Figure 2.14   Block 5: controlling two motors with one switch (S1) (II).

biased, and motor M2 is on. The diodes are mounted near the motors, and their positions must be chosen according to the desired direction that the motor must turn. Diodes 1N4002, or others of the same series, are suitable for motors up to 1 A.

Block 6   Half-Wave Rectifier

When powering a dc motor from an ac power source (e.g., the secondary of a low-voltage transformer), it is necessary to add a *rectifier* to the circuit. The complete block diagram of a half-wave rectifier with filter is shown in Figure 2.15.

The diode conducts only half of the semicycles of the ac power line voltage. The capacitor filters (smooths out) the voltage pulses, keeping the voltage level applied in the motor as constant as possible. The value of the capacitor depends on the current drained by the motor in these applications. A simple rule for motors powered from 3 to 15 V is adopt 1,000 µF of capacitance for each ampere of current. For instance, use a 470 µF (500 µF) for a 500 mA motor. Using this circuit, you can powered your application from such ac supplies as small transformers.

Block 7   Full-Wave Control

The circuit shown in Block 6 uses only half of the semicycles of the ac voltage. If you don't intend to use the other semicycles in another part of the circuit (to control another motor, for instance), it would be better use the configuration shown in Figure 2.16.

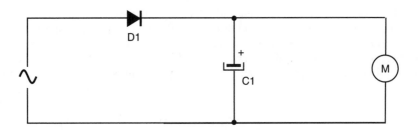

Figure 2.15   Block 6: half-wave rectifier.

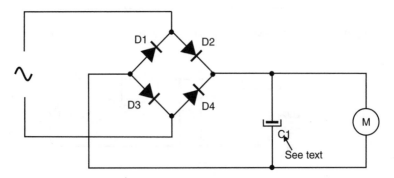

Figure 2.16   Block 7: full-wave control.

The filter capacitor is chosen in the same manner as described in Block 6. This circuit can also be used in dc circuits to keep the direction of a motor independent of the polarity of the voltage source, as shown in Figure 2.17.

In this figure, we show a circuit in which a single switch is used to reverse the dc power supply, reversing the direction of motor M1 but not M2. M2 is wired to the same line, but it runs only in one direction, thanks to the design of this block.

Block 8   Two-Speed Block Using Diodes

Using one diode only causes a voltage drop of about 0.7 V (independent of the amount of current flowing across it). If we want larger voltage drops when controlling motors powered from 6 to 15 V supplies, more than one diode can be used. The diodes must be wired in series, as shown in Figure 2.18.

Each diode causes a voltage drop of about 0.7 V. This means that three diodes will reduce the applied voltage by 2.1 V. Starting from a 6 V power supply, you can power a motor with 6 or 3.9 V in a two-speed system. The diodes are 1N4002, which are suitable for motors up to 1 A.

Block 9   Multi-speed Control Using Diodes

If multiple diodes are wired in series with a motor, the speed reduction factor can be selected by a switch or sensors in the circuit. This is shown in Figure 2.19. The number of diodes used in each section of the circuit determines the reduction step of the system. If you use three diodes in each step, for instance, you can control the speed by varying the voltage in steps of 2 V.

The switch is a 1 pole $\times$ n throw, where n is the number of speeds (switch positions) you want to the control. Note that, in the first position, the motor runs at its maximum speed, as no diode is in the operating part of the circuit.

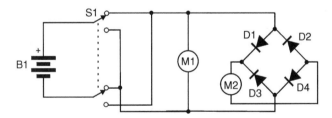

Figure 2.17   Independent control for M1 without affecting the direction of M2.

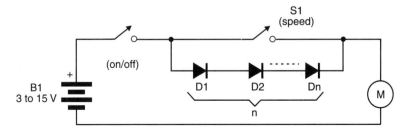

Figure 2.18   Block 8: two-speed block using diodes.

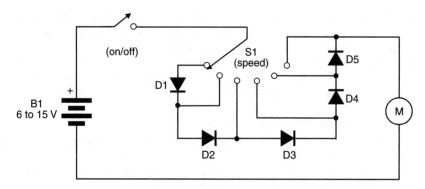

Figure 2.19   Block 9: multi-speed control using diodes.

Block 10   Two-Speed Voltage Change from the Source

The block shown in Figure 2.20 uses the voltage source to control the speed or power of a motor in a robotics or mechatronics application. The power supply is a dual-voltage source as shown in the figure.

By selecting the voltage, it is possible to control the speed or torque of the motor. For instance, we can use a 12 V motor and power it from a 9 + 3 V power supply. With the switch in one position (A), 12 V is sourced to the motor, and in position (B), 9 V is sourced. If we use multiple cells and a switch with one pole and many positions, we can control the applied voltage in steps of 1.5 V.

Block 11   Power Booster

In some of our applications, a motor runs at a speed that is determined by the load. But if, for any reason, the load increases (e.g., the robot encounters an obstacle), the motor may stop. A momentary boost in power may allow the robot to move the obstacle out if its way. Such a boost can be produced applying by a momentary voltage that is higher than nominal. The Block 11, shown by Figure 2.21, can accomplish this.

With S1 not engaged, the power is applied to the motor from battery B2. If S1 is pressed momentarily, the voltage applied to the motor increases by a step determined by B2. The power boost can be enough to help the robot or other device pass

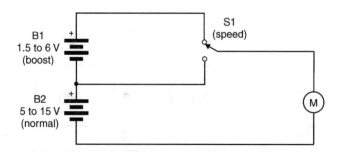

Figure 2.20   Block 10: two-speed operation using two batteries.

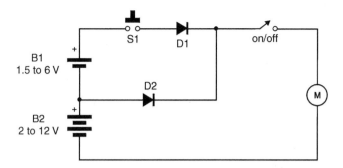

Figure 2.21   Block 11: power booster.

or remove the obstacle. A relay in the robot with a "boost" function in the control can be used in these projects. Sensors, or even a microprocessor, can also be used to control the operation of this booster block. The function of diode D2 is to keep battery B1 from encountering a short circuit when S1 is pressed.

Block 12   Multi-step Power Booster

The Block 11 can be altered to use multiple booster batteries, allowing the operator or the circuit to try different voltage increases. This can be accomplished as shown in Figure 2.22. Remember that the booster batteries shouldn't be connected for long time intervals, since they allow the motor to draw a voltage that is higher than the nominal. For short periods of operation (a few seconds), this is acceptable, but it can cause the motor to overheat over longer periods.

Block 13   Adding Inertia

When you switch on a dc motor, it starts up with maximum power and tends to apply all its torque to the load that must be moved. The effect of this abrupt action is a

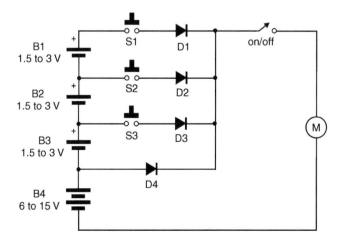

Figure 2.22   Block 12: multi-step power booster.

tendency to produce sudden changes in position, in some cases affecting the equilibrium of the system. For the same reason, when you turn off the motor, the natural tendency is for it to stop immediately, with the same effects. If the motor is used to move a robot, for instance, this abrupt stop can cause it to fall over.

With the block shown in Figure 2.23, we can add some inertia to a dc motor. This means that, when you power up the circuit, the motor produces a "soft" start, because the capacitor must be charged. This is shown by the curves in Figure 2.24.

By the same principle, when the motor is turned off, the energy stored in the capacitor is released, powering down the motor over a few seconds. The larger the capacitor, the more inertia is added.

In a robot, for instance, this means that the machinery doesn't encounter a sudden stop when the motor is turned off; due the inertia, the stop is soft.

You can experiment with capacitors with capacitances between 100 and 4,700 μF, depending the application, when using motors up to 1 A and 15 V.

Block 14   Series–Parallel Switch

An interesting way to boost the power applied to a motor is by using two batteries. When they are connected in parallel, more current is provided, but the voltage is lower. When they are placed in series, more voltage is applied, but the current is lower. This switching process can be accomplished as shown in Figure 2.25. A DPDT switch is used. Of course, this switch can be replaced by the contacts of a relay.

An important factor to observe in this circuit is that the batteries must have the same voltage. If they are different, the current tends to flow through the lower voltage battery when they are in parallel. As there is nothing in the circuit to limit this current, the batteries can be destroyed. Keep in mind that the same block can be used to connect any other voltage source (or signal) in series or parallel.

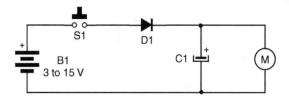

Figure 2.23   Block 13: adding inertia.

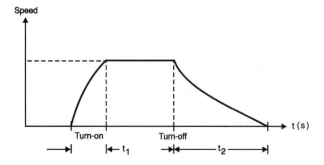

Figure 2.24   Adding inertia with a capacitor.

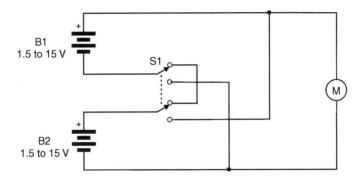

Figure 2.25   Block 14: series–parallel switch.

Block 15   Motors in Series and Parallel

Another important application for the series–parallel switch is in the control of two motors, as shown by the block in Figure 2.26. The motors used in this application must have the same characteristics (voltage and current).

When they are wired in parallel, each is powered by the main power supply voltage. When they are wired in series, the voltage is divided by two, and each one receives half of the supply voltage while running in a low-speed mode.

But the circuit does not always behave in this manner. If one motor is loaded more heavily than the other, the voltage will no longer be divided by two. The heavily loaded motor will be powered by a lower voltage, and the other will run with higher voltage. In other words, in this circuit, the voltage across each motor depends on the load.

Block 16   LED Direction Indicator

The block shown by Figure 2.27 can be used with dc motors powered from 3 V supplies. Which LED is turned on depends on the direction of the current flowing across the circuit. You can use a green LED to indicate forward and a yellow LED to indi-

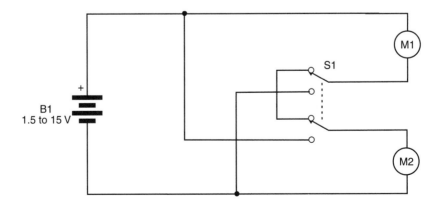

Figure 2.26   Block 15: motors in series–parallel.

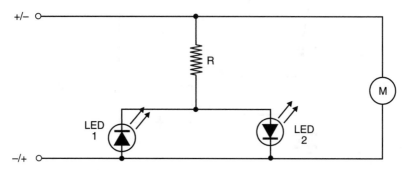

Figure 2.27   Block 16: LED direction indicator.

cate reverse. The value of resistor R depends on the voltage used to power the circuit, according to the following table:

Voltage	R
3 V	47 Ω
4.5 V	150 Ω
6 V	390 Ω
9 V	680 Ω
12 V	1.2 kΩ
15 V	1.8 kΩ
18 V	2.7 kΩ
24 V	3.3 kΩ
36 V	4.7 kΩ
48 V	6.8 kΩ

If the motor operates in a range of voltages via the use of a speed control, use the appropriate resistor for the highest applied voltage. All the resistors are 1/4 W × 20%.

Block 17   Current Indicator

The LED in the block shown by Figure 2.28 glows when current is flowing across the motor. The brightness is largely independent of the amount of current flowing.

The three diodes cause a voltage drop of about 2.1 V. The red LED needs about 1.6 V to be forward biased and lit up. If you are using another type of LED (such as yellow, which needs 1.8 V, or green, which needs 2.1 V), you may need to add one or two diodes to the circuit to the voltage drop. In any case, remember that the diodes reduce the voltage that will be applied to the motor.

Block 18   Adding Sound (I)

The switching process that occurs inside a dc motor produces a sound at the motor's operational frequency. If a small loudspeaker is wired in series with a dc motor, the voltage generated by the switching process can be treated as a signal and reproduced. This is the aim of the block shown in Figure 2.29.

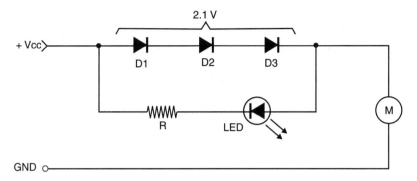

Figure 2.28 Block 17: current indicator.

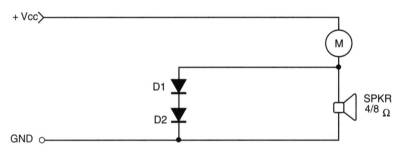

Figure 2.29 Block 18: adding sound (I).

One or two diodes placed in parallel with the loudspeaker avoid the voltage drop in this component (because of its impedance), keeping the torque and speed constant. Any small loudspeaker (2 to 4 inches and 4 to 8 Ω) can be used here. This block uses two diodes and is indicated for the current flowing only in one direction.

Block 19   Adding Sound (II)

Small piezoelectric transducers placed in parallel with dc motors will reproduce the signal generated in the switching process as shown by Figure 2.39. The diode is necessary, as the buzzer is powered from the reversed voltage generated in the switching process by the inductance produced by the motor's coils.

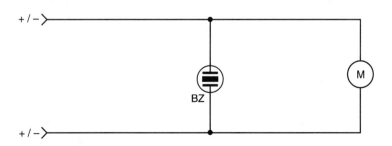

Figure 2.30 Block 19: adding sound (II).

Block 20   Two-Wire × Two-Motor Control

The circuit illustrated by Figure 2.31 combines Blocks 1, 2, and 7 in a single complex application for robotics and mechatronics. Block 1 is the on/off control for the two motors (M1 and M2). Switch S2 controls the direction of the current in the remote circuit where M1 and M2 are placed.

In this configuration, M1 is sensitive to current flow and reverses its direction when the current flow changes. But M2, due the presence of Block 7, is not sensitive to current flow, and it doesn't reverse its direction when the flow reverses.

One application for this circuit is a robot in which a fan is powered from the same circuit that powers the propulsion motor. The fan motor is M2. If the current in the circuit is reversed, the fan continues to run in the same direction.

2.5   Suggested Projects

The simplest way to add directional control for a robot is using two motors as shown by Figure 2.32. If the two motors are on, the robot goes in a straight line. If one motor is on and the other off, the robot turns in the direction of the motor that is off. The same type of control can be provided if the motors runs at different speeds.

Simple mechanical systems such as shown by Figure 2.33 can also be assembled with a single dc motor coupled to a gearbox. The up and down movement and speed can be controlled directly by the motor.

Starting from these concepts, some projects can be suggested:

- Draw a control circuit for a small robot that can walk in all directions, using single switches for this task.
- Design an experimental elevator controlled by switches.

2.6   Additional Information

Small HE (high-efficiency) and LE (low-efficiency) dc motors and gearboxes can be purchased from many component dealers. The most popular is Radio Shack, which

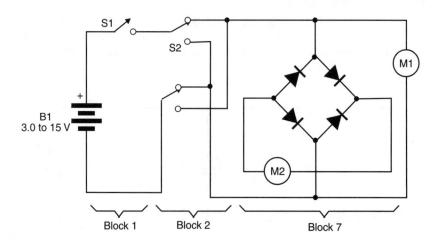

Figure 2.31   Block 20: two-wire × two-motor control using three blocks.

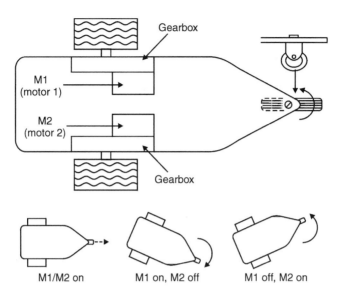

Figure 2.32   Directional system using two motors.

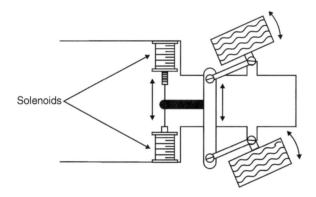

Figure 2.33   Direction control using solenoids.

can provide the designer with many small motors and gear kits such as those listed below:

- 1.5 to 3 VDC, 8300 rpm, part no. 273-223
- High-speed, 12 V, 15,200 rpm at no load, part no. 273-255
- Super-speed, 9 to 18 V, 24,000 rpm at no load, part no. 273-256
- Gear kit, two output shafts for different speeds and torque, ideal for use with motor 273-223, part no. RSU 11903135

2.7   Review Questions

1. What determines the rotational direction of a dc motor?

2. How can we control the speed of a dc motor?
3. What are gearboxes?
4. What is a silicon diode?
5. What is the voltage drop across two silicon diodes that are forward biased? Use the 1N4002 for this solution.
6. If two batteries are placed in series, what happens to the voltage provided by them?

3

Controlling Motors, Relays, and Solenoids with Transistors

3.1 Purpose

In this chapter, we will show how transistors (bipolar and power FETs) can be used to control loads such as motors, solenoids, and relays. The practical blocks can be used in projects involving both motion and circuit control.

3.2 Theory

In Chapter 2, we examined how we can control a dc motor by changing the direction of current flow and the applied voltage. Two new elements can be added to our projects to allow not only the control of motors but also other loads such as solenoids, electromagnets, lamps, and electronic circuits. The following blocks will show the reader how to use two new elements: the relay and the transistor.

3.2.1 The Relay

Relays are electromechanical switches. They are basically formed by a coil and one or more pair of contacts as shown by Figure 3.1. When a voltage is applied to the coil, the current flow creates a magnetic field that attracts the contacts and closes them. If the current across the coil is cut, the magnetic field disappears, and the con-

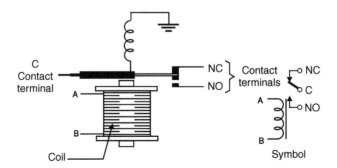

Figure 3.1 Relay structure and symbol.

tacts open. An external circuit controlled by these contacts can be turned on and off by the current across the coil.

Two important properties can be noted in this arrangement:

1. The controlled circuit is completely isolated from the control circuit.
2. We can apply low voltage and low current to the relay's coil to control high-voltage and/or high-current circuits.

Relays can be obtained in different sizes and with different electrical characteristics as appropriate for a given application.

When using relays in these projects, we must observe the following characteristics:

- The *nominal* or *coil* voltage. This is the voltage that, when applied to the coil, makes the contacts close. In practice, a relay can close its contacts with voltages lower than indicated, and keep them closed even when the voltage falls below the nominal value, demonstrating the characteristic of *hysteresis*. Relays with nominal voltages between 3 and 48 V are common in robotics and mechatronics applications.
- The coil current. This is the current that flows across the coil when the nominal voltage is applied. Currents between 20 and 100 mA are common in relays.
- The coil resistance. This resistance can be found easily by dividing the nominal voltage by the coil current. Values between 50 and 500 Ω are common.
- Contact current. This is the maximum current that can be controlled by a relay. Typical values are in the range of 1 to 10 A.

How Relays Are Used.   Figure 3.2 shows the symbol and the appearance of common relays. A relay can be considered as a switch that is operated by an electric signal. Figure 3.2a shows an SPST relay, and 3.2b shows a DPDT relay. Other types, with multiple contacts, can be obtained.

3.2.2 Basic Blocks Using Relays

The following are blocks that use relays and some passive components such as diodes, resistors, and capacitors. More complex applications, including other elements coupled to the relays, will discussed in later parts of this book.

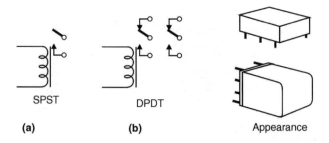

SPST   DPDT

(a)   (b)   Appearance

Figure 3.2   Types of relays.

Block 21   Turning a Load On or Off

An important detail to observe in these relays is that they can have contacts that are normally open (NO) or normally closed (NC) as shown in the symbols. The simplest way to control a load from a relay is to use one of these contacts and the common (C) as shown in Figure 3.3.

If the NO and C contacts are used, the load is powered on when the relay closes the contacts (when its coil is energized). If the NC and C contacts are used, the load is turned off when the relay is energized. This means that you can use the relay to turn loads on or off. Remember that the contacts are closed only while current is flowing through the coil.

Block 22   Current Reversion

Another use for a relay is to reverse current, as shown in Chapter 2, when using the DPDT switch. The relay is used as shown by Figure 3.4.

Block 23   Series–Parallel Switching (I)

This is the same application described in Block 14 but with the switch replaced by a relay. The block shown in Figure 3.5 functions as follows:

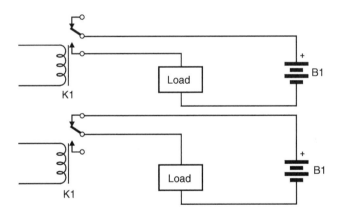

Figure 3.3   Block 21: turning a load on or off.

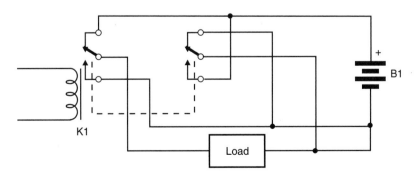

Figure 3.4   Block 22: reversing current with a relay.

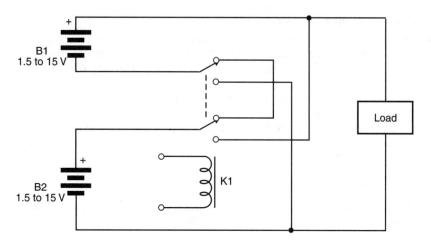

Figure 3.5   Block 23: series–parallel switching (I).

- When the relay is off (no current flowing through the coil), the batteries are wired in series. When the relay is energized, the batteries become wired in parallel. The batteries can be replaced by other circuits (a signal source, for instance).
- Observe that the voltage applied to the relay for the control of the loads can be different from the voltage of the batteries.

Block 24   Series–Parallel Switching (II)

The circuit shown by Figure 3.6 switches two loads, connecting them in series or parallel to a power supply. When the relay is on, the loads are connected to the voltage source in series. When the relay is off, the loads are connected in parallel. The

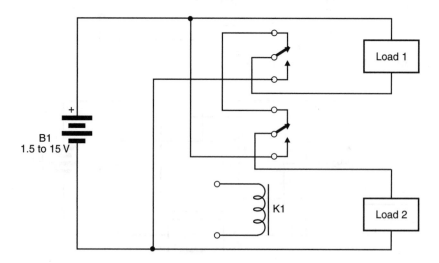

Figure 3.6   Block 24: series–parallel switching (II).

circuit can be used to boost dc motors in robotics and mechatronics devices. The same circuit can be used in an inverse mode, switching the loads in parallel when the relay is on.

Block 25   Timed Relay

A brief voltage pulse is used to energize the relay. However, the capacitor keeps the relay energized for a period of time after the pulse is gone, depending on the capacitor value. In the block shown by Figure 3.7, the pulse is applied by a momentary switch, but any sensor can be used in robotics or mechatronics applications.

Examples of sensors include reed switches and microswitches that can be activated by very short pulses. This is useful when circuit to be powered up needs to remain on for a longer period of time to work efficiently.

The proper value of the capacitor depends on the time the relay remains on, and also on the relay's resistance. Typical values for relays between 50 and 1,000 Ω are in the range of 100 to 4,700 μF. Observe that the relay will be on for a time interval that begins when S1 is pressed and ends when the voltage falls to the minimum level that will keep the contacts closed (not when it falls to 0 V).

Block 26   Short-Action Relay

Using the circuit shown in Figure 3.8, the relay is closed briefly when S1 is pressed (or a sensor is activated). The time interval in which the relay is on depends on the capacitor value and the resistance of the relay's coil. The time constant of the circuit is measured from the instant in which S1 is pressed to the point at which the voltage

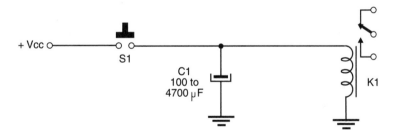

Figure 3.7   Block 25: timed relay.

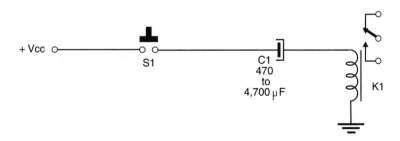

Figure 3.8   Block 26: short-action relay.

in the coil falls to the minimum that keeps the contacts closed *(holding voltage)*. Relays between 470 and 4,700 µF can be used in common applications.

Block 27   Flasher

The relay will open and close its contacts at a frequency that is determined by its coil resistance and the value of the capacitor. A lamp will be turned on and off according this frequency.

The circuit shown in Figure 3.9 can use relays with coils between 50 and 500 Ω, and the capacitor is specified in a range between 100 and 4,700 µF. This configuration can be used in signaling systems for robots or even to flash decorative lamps.

Block 28   Passive Buzzer

An interesting configuration shown by Figure 3.10 can be used to generate a continuous tone in a small loudspeaker or even in a piezoeletric transducer. The frequency is determined by the time constant given by the inductance of the relay's coil and the capacitance, C. Capacitors rated between 0.047 and 470 nF can be used experimentally in a particular application. Any small loudspeaker can be wired in series with the relay. Piezoelectric buzzers must be wired in parallel with the coil.

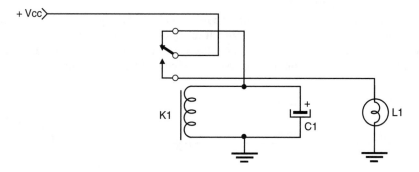

Figure 3.9   Block 27: flasher.

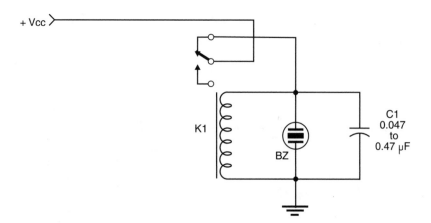

Figure 3.10   Block 28: passive buzzer.

Block 29   Locked Relay

When S1 is momentarily pressed, the relay closes its contacts and remains in this state until the supply voltage is cut off. S1 can be any sensor, microswitch, or other momentary contact switch as suitable for the application. If a DPDT relay is installed, the second pair of contacts can be used to control any external circuit, as suggested in Figure 3.11.

To turn off the circuit, you can simply cut off the power supply voltage, but there are other possible approaches. One of them is to create a momentary short using a switch in parallel with the coil. Another solution is given in the next block.

3.3   The Transistor as a Switch

Relays are not the only components that can be used to control a load with an electric signal. The transistor is also applicable to this function, and there are many advantages do doing so. They are ideal for many applications in robotics and mechatronics. The principal advantages are as follows:

- Low cost
- Small size
- High speed
- High sensitivity

There are two basic types of transistors: bipolar and field effect. Let's see how they can be used as switches.

3.3.1   Bipolar Transistors

There are two types of bipolar transistors, categorized according to the internal structure of the semiconductor material from which they are manufactured, as shown in Figure 3.12. The basic difference between the two types is the direction of the current flow while they are functioning. As shown in Figure 3.13, the NPN transistor conducts a large current between collector and emitter when the base is made positive with respect to the emitter.

In a typical transistor, the current between collector and emitter can be 100 times higher than the current applied to the base, meaning that this devices can produce gains of 100× or more. In the operation of an NPN transistor, the collector

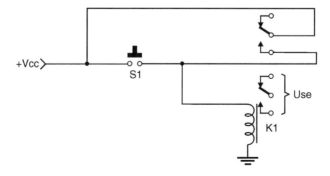

Figure 3.11   Block 29: locked relay.

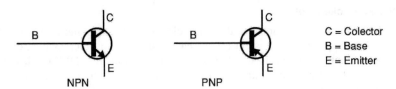

C = Colector
B = Base
E = Emitter

NPN   PNP

Figure 3.12   Types of bipolar transistors.

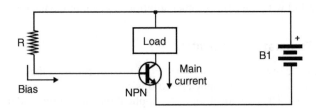

Figure 3.13   Operation of an NPN transistor.

must be connected to the positive pole, and the emitter to the negative pole, of the voltage source.

On the other hand, the PNP transistor conduces a large current between the emitter and collector (the reverse direction as compared with an NPN transistor) when the base is made negative with respect to the emitter, as shown by the Figure 3.14. The voltage source must have the positive side connected to the emitter and the negative to the collector. The load to be controlled by a transistor can be installed between the collector and the power supply or between the emitter and power supply as shown by Figure 3.15.

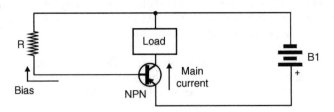

Figure 3.14   Operation of a PNP transistor.

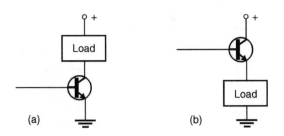

Figure 3.15   Wiring the loads to an NPN transistor.

The main characteristics to be considered in a bipolar transistor, when using it in a project design, are the following:

- V_{ce}. This is the maximum voltage that can be applied between collector and emitter without burning up the transistor. Typical values are between 20 and 500 V.
- I_c. This is the maximum current that can flow through the collector terminal; it is the maximum current that can be controlled by the transistor. Typical values are between 100 mA and 10 A. Transistors that can control currents above 500 mA are termed *power transistors* and normally must be mounted on heat sinks.
- P_d. This refers to maximum power dissipation, i.e., the maximum amount of power that can be converted to heat by the transistor. Transistors with a P_d of 1 W or more are also classified as *power transistors* and must be mounted on heatsinks.

In robotics and mechatronics circuits, a bipolar transistor can be used to perform many tasks. One is just turning a load on and off by applying a small current to the base. The load can be a relay, motor, lamp, or any other electronic circuit, as we will see in the following blocks.

3.3.2   Field Effect Transistor or FETs

Field effect transistors are produced in four basic types, as shown by Figure 3.16. In 3.16a, we see the junction FET (J-FET), and in 3.16b, we represent the two types of metal-oxide semiconductor field effect transistors (MOSFETs).

The operating principle of these transistors is the same: the current between drain and source can be controlled by a voltage applied to the gate. The direction of the current and the polarity of the voltage applied to the gate (G) depends on the type.

The N-channel conducts the current when the gate is made positive with respect to the source. Figure 3.17 shows an application in which the load is installed between the power supply and the drain (d) of the transistor. FETs can be used for the same functions as bipolar transistors, as we will see in the blocks presented in this section.

The main characteristics to be observed in FETs are the following:

- V_{ds}. This is the maximum voltage that can be applied between the drain and source. Typical values are in the range between 50 and 1000 V.

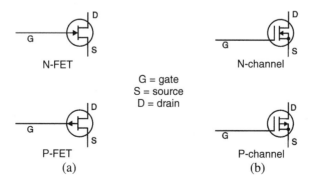

Figure 3.16   Field effect transistor types.

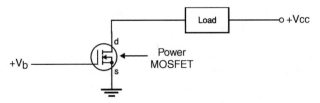

Figure 3.17   The power MOSFET.

- I_d. This is the maximum current across the device. Values between 50 mA and 100 A are common. Special transistors, referred to as power FETs or power-MOSFETs, can control currents above 1 A. These are power transistors that must be mounted on heat sinks.
- P_d. This is the maximum amount of power that the transistor can convert to heat when operating.

3.4   Practical Blocks

The following basic circuits using bipolar and power-MOSFET transistors can be useful in robotics and mechatronics.

Block 30   Switching with an NPN Transistor

The block shown by Figure 3.18 shows how a bipolar NPN transistor can be used to control a load. The load can be a relay, motor, solenoid, or lamp. The diode (1N4002 or equivalent) is necessary if inductive loads such as motors and solenoids are driven.

If you are controlling a relay (50 to 500 Ω, up to 100 mA) or any load up to 100 mA, the transistor can be a BC547, BC548, or equivalent. The BD135 can control loads up to 1 A, and the TIP31 up to 2 A. These transistors are mounted on a heat sink.

R1 is used to limit the base current. The circuit will turn on when a voltage between 1 and 12 V is applied to the input. A current of about 100 μA is necessary to drive this circuit when using the BC548.

In the same figure, we show how a switch can be used to control this circuit. Sensors such as LDRs can also be used, replacing the switch.

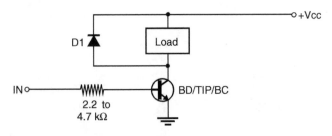

Figure 3.18   Block 30: switching with an NPN transistor.

Block 31   Switching with a PNP Transistor

The circuit shown by Figure 3.19 can drive the same loads as the previous blocks. The PNP transistor used for loads up to 100 mA can be the BC558. For loads up to 1 A, use the BD136 or BD138, and for loads up to 3 A, the TIP32 or even the TIP42.

The load is turned on when the input is connected to ground, or a negative voltage is applied. With an input current of about 100 µA, loads up to 100 mA can be driven, depending on the transistor gain. This sensitivity is enough to allow the use of resistive sensors as LDRs. The diode is necessary when driving inductive loads.

Block 32   Using an NPN Darlington Pair

Two directly coupled NPN transistors can increase the gain of a driver stage as shown by the block in Figure 3.20. The gain of the stage is the gain of Q1 times the gain of Q2. With a few milliamperes, it is possible to drive high-current loads such as motors, solenoids, and relays.

Q1 can be any NPN general-purpose silicon transistor such as the BC548 or 2N2222. Q2 depends on the current of the load. For sensitive relays (up to 50 mA), the BC548 or 2N2222 can be used. When driving directly loads up to 1 A, use the BD135 or BD137 and, when driving larger loads up to 3 A, use the TIP31 or TIP41 mounted on a heat sink. Diode D1 is necessary when working with inductive loads.

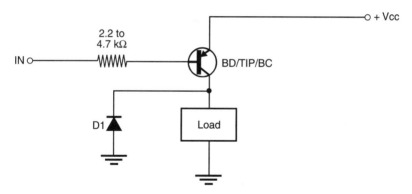

Figure 3.19   Block 31: switching with a PNP transistor.

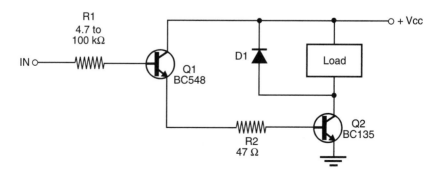

Figure 3.20   Block 32: using NPN Darlington pairs.

The input resistor depends on the signal source. When the input receives a positive voltage between 2 and 12 V, the load is turned on. A few microamperes are necessary to drive the circuit. An LDR can be used to excite this stage.

Block 33   Using a PNP Darlington Pair

The equivalent configuration of the previous block, using PNP transistors, is shown by Figure 3.21. For loads up to 50 mA, use the BC558 or any general-purpose silicon PNP transistor. For loads up 1 A, use the BD136 or BD138 and, for loads up to 3 A, the TIP32 or TIP42.

The circuit is driven when the input is connected to ground by a switch, any resistive sensor, or an external circuit. A few microamperes are necessary to turn on the load. The diode is also required when driving inductive loads.

Block 34   Using a Darlington NPN Transistor

Instead of using two separated transistors to form a Darlington pair, it is possible to use a device that contains two transistors wired as a Darlington pair. These devices are called Darlington transistors, and they can be found in NPN rather than PNP form. Figure 3.22 shows how an NPN Darlington transistor can be used to drive a load as shown in the previous blocks.

The transistor shown is a TIP111 that can control currents up to 1 A. It has a current gain as high as 1,000. This means that an 1 μA current in the input can cause a 1 A current in the load. As in the equivalent NPN block, the load is on when a positive voltage is applied to the input.

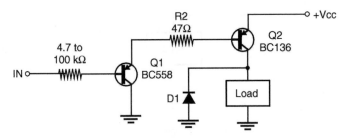

Figure 3.21   Block 33: using a PNP Darlington pair.

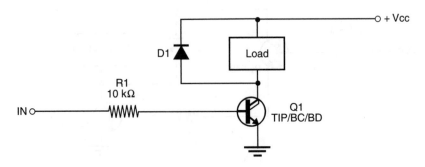

Figure 3.22   Block 34: using a Darlington NPN transistor.

Block 35   Using a Darlington PNP Transistor

The equivalent block for the PNP Darlington transistor is shown by Figure 3.23. The characteristics are the same as those of the previous block except that the circuit is turned on when the input is connected to ground.

Block 36   Complementary Driver (I)

In the block shown by Figure 3.24, we have two transistors (one NPN and the other PNP) directly coupled to form a high-gain stage for driving loads as in the previous blocks. Q1 is an NPN transistor. Any general-purpose NPN silicon transistor such as the BC548, 2N3904, 2N2222, or others can be used for this task. Q2 depends on the current across the load. For loads up to 50 mA, such as small relays, magnets, solenoids, lamps, and motors, a BC558 can be used. For large currents, you can use the BD136 (1 A) or TIP32 (3 A). A few microamperes in the input are necessary to drive the loads. The load is on when the input is positive.

Block 37   Complementary Driver (II)

This block (Figure 3.25) also uses complementary transistor, but the input (Q1) is a PNP, and the output (Q2) is a PNP. For Q1, you can use the BC558 or any equivalent. For Q2, you can use the BD135 (1 A) or the TIP32/TIP42 (up to 3 A). The load is on when the input is low (to ground). Only few milliamperes are necessary to drive the loads, which allows the use of resistive sensors such as LDRs.

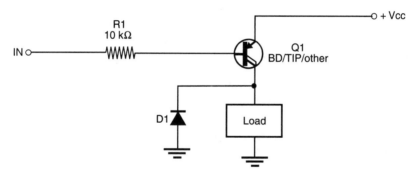

Figure 3.23   Block 35: using a Darlington PNP transistor.

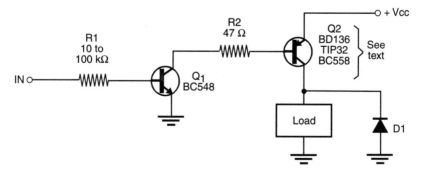

Figure 3.24   Block 36: Complementary driver (I).

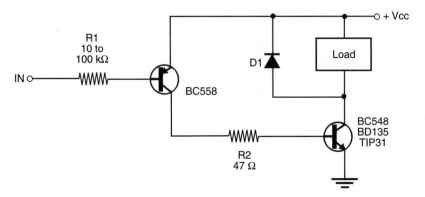

Figure 3.25 Block 37: complementary driver (II).

Block 38   Power FET Driver

The block shown in Figure 3.26 can drive loads up to several amperes from a voltage source that is typically between 2 and 12 V, depending on the power supply voltage. When the input is positive, the power FET conducts, and the load is powered on. The most important characteristic of this block is the low R_{ds} on. R_{ds} is the resistance between the drain and source of a power FET. This resistance can fall to values as low as fraction of an ohm (0.01 Ω is found in some devices). This means that practically no power is dissipated in the transistor, which can control high currents easily.

However, the reader must bear in mind that low R_{ds} on is valid only when high control voltages are used. When operating with low-voltage sources, the power FET has the same characteristics of the bipolar transistor when conducing the current, and some power is converted into heat. Another important characteristic of this circuit is its very high input impedance.

3.5   More Blocks Using Relays and Transistors

When using a transistor to drive a relay, some specific configurations can be suggested to the robotics and mechatronics designer. The transistor can add sensitivity

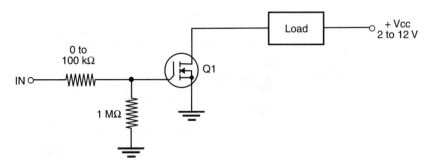

Figure 3.26 Block 38: power FET driver.

to a relay and, in addition, some special characteristics that, if used in a project, can be very interesting for specific tasks. The following blocks are variations of the previous and use the transistor as a switch as described in theory. Each block can be changed, adapted, or used a starting point of other configurations.

Block 39   Delayed Relay

In projects involving certain kinds of mechanisms, the designer needs to employ a circuit that closes a relay not in the instant that the power is switched on but after a delay of several seconds. For example, a robot may need to complete a particular movement before a stop mechanism is activated. The block shown in Figure 3.27 can be used for this task.

The relay can be any type with a voltage rating between 6 and 15 V and a coil drain of not more than 50 mA. The transistor is a 2N2222, BC548, or any general-purpose NPN transistor.

The delay is given by the time constant RC and can be adjusted to assume values from 0.5 to about 10 seconds. Larger intervals are critical, as they depend on capacitor leakage and transistor gain. For large intervals, use other blocks shown in this book.

Block 40   Timed Relay

Another block you might need for a project is one that turns a circuit off several seconds after a sensor or switch is opened. The circuit shown in Figure 3.28 can be used for this task.

When S1 is activated, capacitor C1 is charged, and the relay is turned on. When switch S1 is released, the capacitor discharges through R1, and the transistor keeps the load activated for a time. The length of time depends on C and R values and can be adjusted from 0.5 to about 10 seconds. It is not recommended for higher intervals due the leakage of the capacitor. Other configurations for large intervals are provided elsewhere.

Block 41   Long-Interval Delayed Relay

This block is used in the same manner as Block 39, but it can provide delay intervals of up to 10 minutes, depending on the values of C and R. Figure 3.29 shows the complete diagram of the block.

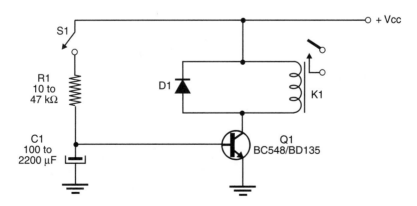

Figure 3.27   Block 39: delayed relay.

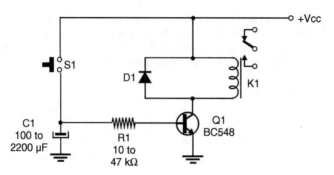

Figure 3.28   Block 40: timed relay.

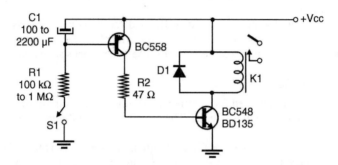

Figure 3.29   Block 41: long-interval delayed relay.

The relay can be any type with a coil current up to 50 mA, and the circuit can be powered from 6 to 15 V according to the relay and the applications. Q1 is any general-purpose NPN transistor such as the BC548 or 2N2222, and Q1 is a general-purpose PNP transistor such as the BC558.

Block 42   Long-Interval Timed Relay

This circuit is the equivalent of Block 40 but for larger time intervals. Time intervals from several seconds to a few minutes can be achieved with the circuit shown in Figure 3.30.

When S1 is activated (S1 can be a switch or any sensor), capacitor C charges, activating the relay. When the switch is released, the capacitor discharges across the circuit, keeping the relay on. The relay will remain on for a time period depending on the values of C and R.

Block 43   Delayed Relay Using Power FET

The high input impedance of a power MOSFET can help the designer to achieve wide time intervals in a delayed relay, as shown by Figure 3.31. When the sensor or switch S1 is closed, capacitor C1 begins to charge across R1 until the voltage in the gate becomes high enough to turn it on. The high impedance of the FET's input makes possible the use of high-value resistors in this circuit, meaning longer time intervals. With a 1,000 µF capacitor and a 1,000,000 Ω resistor, the time delay can be extended to several minutes.

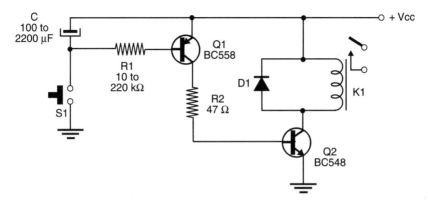

Figure 3.30   Block 42: long-interval timed relay.

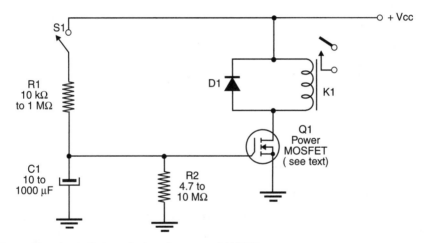

Figure 3.31   Block 43: delayed relay using a power MOSFET.

This circuit, like the other blocks, can be used to control other devices, but the designer must keep in mind that the voltage in the load when the circuit is activated (or deactivated) changes slowly. This is an important point to consider when the blocks are used to control solenoids, motors, or other circuits.

Block 44   Holding Circuit

This circuit can be used to keep a relay closed for very long time intervals. We can think of it as a lock or holding circuit that is suitable for many applications in robotics and mechatronics. The circuit, for instance, can be used to stall a system or a robot when a particular sensor is activated, disabling the power supply or the propulsion system. The circuit uses a power MOSFET and can drive not only relays but other loads such as motors, solenoids, and circuits. The circuit is shown in Figure 3.32.

When S1 is activated or a positive control pulse is applied to the input IN, capacitor C1 is charged. Due the very high input impedance of the power MOSFET input,

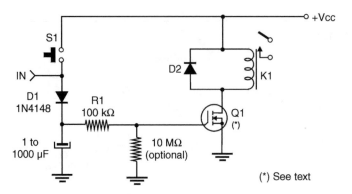

Figure 3.32   Block 44: holding circuit.

the capacitor maintains its charge (D1 is used if an external circuit is plugged to the input, avoiding the discharge of C1). The circuit will remain activated as long as the capacitor keeps its charge. A second switch can be placed in parallel with the capacitor to disarm the circuit.

An important point to consider when assembling this circuit is that capacitor C1 must be of high quality, without leakage. Polyester types from 1 to 10 µF are recommended.

We also must consider that humidity in the air can contribute to capacitor discharge. Keep all wires between the capacitor and gate short in this application to avoid charge leakages into the air.

Any power FET can be used in this application, and the relay or load depends only on the voltage of the power supply.

3.5.1   Working with the Blocks

Using the blocks shown in this chapter, and other blocks described in Chapter 2, we can come up with some interesting suggestions for projects in mechatronics and robotics. The following are only three of many ideas that can be explored by designers. These blocks can be altered, expanded, or even ganged to other blocks to form complex projects. It is up to the reader's imagination to arrange all the blocks in the infinite combinations that they permit.

Block 45   Reversing a Motor for a Few Seconds

The circuit shown in Figure 3.33 can be used in a project involving a small moving robot with some "intelligence." X1 is a movement sensor, used as an "antenna" to detect obstacles. When the sensor is momentarily activated in the presence of an obstacle, the motor is reversed for a few seconds (as determined by the RC circuit). This is enough to move it back by several centimeters, after which the motor is again reversed. If some kind of mechanism is used to change the robot's direction, it can now proceed forward in other direction, attempting to find a clear passage.

As the reader can see, this applications is based in the Block 40, giving several seconds of reverse movement. The latch given by Block 44 can also be adapted to the circuit, making it continue until an external action changes this condition. The component values shown in the circuit are for typical applications and can be altered as required.

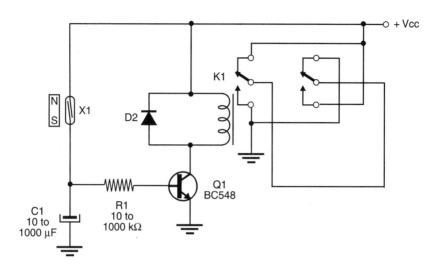

Figure 3.33 Block 45: reversing a motor for a few seconds.

Block 46 Reversing and Stopping a Motor

This application circuit, based in Blocks 40 and 43, can be used in an intelligent robot. The circuit shown in Figure 3.34, when activated by a sensor (as in the previous application), closes the relay, reversing the propulsion system. But, at the same time, C2 charges by R2 until transistor Q1 becomes conductive, closing the second relay. At this moment, the power supply voltage is cut, and the entire system is stalled.

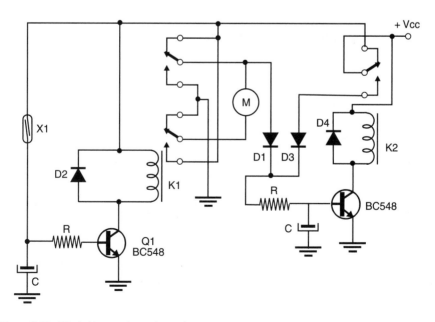

Figure 3.34 Block 46: reversing and stopping a motor.

This configuration can be used to reverse the movement of a robot for a few seconds if it encounters an obstacle. Once it moves several centimeters away from the obstacle (to a secure position), it stops.

The values of R1, R2, C1, and C2, which determine the timing of the circuit, must be established experimentally to suit the application. The values given in the diagram are typical for applications powered from supplies of between 6 and 12 V.

Block 47   Reversing a Motor and Reducing Its Speed

The interesting block shown in Figure 3.35 can be used in several projects involving mechatronics and robotics. If S1, S2, and S3 (sensors) are closed momentarily, relay K1 reverses the direction of the motor. But if the activated sensor is S1, the relay K2 is also activated, and the motor reverses direction and reduces its speed.

The reduction factor is given by the number of diodes used from D4 to D7. In this case, four diodes cause a voltage drop of about 2.8 V in the voltage applied to the motor. The circuit is based on Blocks 35 and 39. The circuit can adapt other blocks for the same application.

3.6   Suggested Projects

The use of relays, bipolar transistors, and power FETs in control and time circuits offers many possibilities to the robotics and mechatronics designer. The following are some ideas for projects using the blocks we have presented so far. Grab your tools and try to create some of these working robots or mechatronics devices:

- Design a circuit that can control the movement of two motors in a robot, using sensors with switches.
- Design an automatic elevator that can stop at each floor for a few seconds. Use switches or reed switches and magnets as sensors.
- Create an automatic door that senses the presence of a person using a switch placed under the carpet. Design it to make some "intelligent decisions" such as not closing if a person is passing through it, etc.

3.7   Additional Information

3.7.1   Information about Relays

Radio Shack has some popular relays that are suitable for all the applications described in this book. We selected some of their relays having the right characteristics.

- Mini SPDT relay for 2 A currents. This relay needs only 18 mA to be activated and operates in the range voltage range of 7 to 9 V. The Radio Shack part number is 275-005.
- Mini SPDT relay for 10 A currents. This relay has a 12 V × 30 mA coil (400 Ω). The part number is 275-248.
- Mini DPDT relay for 5 A. This relay needs 60 mA from a 12 V power supply to close its contacts. The part number is 275-249.

Figure 3.36 shows the pinout for these relays.

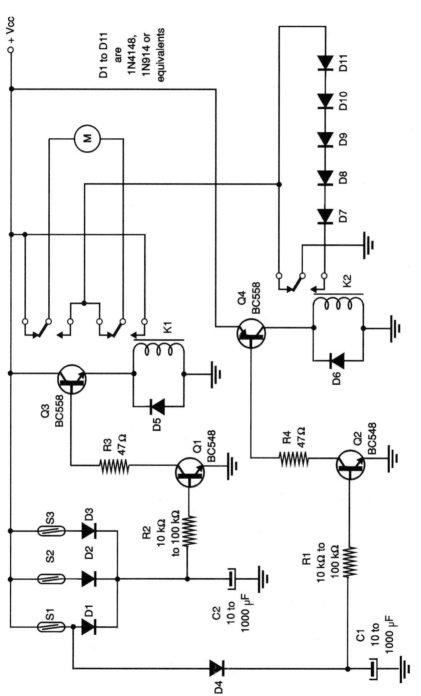

Figure 3.35　Block 47: reversing a motor and reducing its speed.

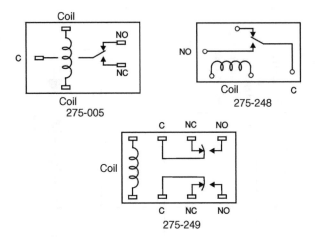

Figure 3.36 Pinouts for some Radio Shack relays.

3.7.2 Information about Transistors

The following tables describe some transistors that are suitable for use as switches in the previously described blocks.

NPN

Type	Voltage (V_{ce})	Current (I_c)	Gain (hFE)
BD135	45 V	1.5 A	40–250
BD137	60 V	1.5 A	40–250
BD139	80 V	1.5 A	40–250
TIP31	40 V	3 A	10–50
TIP41	40 V	6 A	15–75

PNP

Type	Voltage (V_{ce})	Current (I_c)	Gain (hFE)
BD136	45 V	1.5 A	40–250
BD138	60 V	1.5 A	40–250
BD140	80 V	1.5 A	40–250
TIP32	40 V	3 A	10–50
TIP42	40 V	6 A	15–75

Darlington NPN

Type	Voltage (Vce)	Current (Ic)	Gain (hFE)
BD331/333/335	60/80/100 V	6 A	750
TIP110/111/112	60/80/100 V	1.25 A	500
TIP120/121/122	60/80/100 V	3 A	1,000
TIP140/141/142	60/80/100	5 A	1,000

Darlington PNP

Type	Voltage (Vce)	Current (Ic)	Gain (hFE)
BD336/338/340	60/80/100 V	6 A	750
TIP115/116/117	60/80/100 V	1.25 A	500
TIP125/126/127	60/80/100 V	3 A	1,000
TIP145/146/147	60/80/100 V	5 A	1,000

Power FETs (P-Channel)

Type	Vds (max)	Id (max)
IRF511	60 V	4 A
IRF522	100 V	7 A
IRF540	100 V	27 A
IRF611	150 V	2.5 A
IRF620	200 V	5 A
IRF630	200 V	9 A
IRF640	200 V	18 A

3.8   Review Questions

1. What are the main advantages of using relays to switch electric loads?
2. What is the difference between an NPN and a PNP transistor?
3. The bipolar transistor is a typical current amplifier. On the other hand, the power MOSFET is a typical _____ amplifier.
4. Is the impedance a power MOSFET high or low?
5. To turn on an NPN transistor, what is the polarity of the voltage applied to the base?
6. What is the R_{ds} of a power MOSFET?
7. What is the gain of a bipolar transistor?

4

H-Bridges

4.1 Purpose

In this chapter, we will see how the half and full H-bridge can be used to control dc motors. H-bridges are very important to robotics and mechatronics designers, as they simplify circuitry, allowing the control of dc motors directly from electrical signals without the need for relays or other mechanical parts that we described in previous chapters. The purpose of this chapter is to show how the H-bridge works and to give to designers practical blocks and ideas about how to use it. These circuits can also be used in the control of other polarity-sensitive loads such as solenoids and electromagnets.

4.2 Theory

Reversing a motor is a simple task using a relay with DPDT contacts, as we observed with Block 2. But the use of relays presents some inconveniences: relays are relatively expensive, some are large and heavy, and they use mechanical parts that can fail. In addition, it can be difficult to find models with the desired characteristics.

The use of transistors as switches, as shown in Block 3, offers the designer new resources for controlling dc loads. But, unfortunately, reversing a motor in this way requires a DPDT switch, whereas a transistor operating as a switch is an SPST unit. How do we solve this problem?

The solution is provided here. Two or four transistors can be installed together in some special configurations, allowing the control of a dc motor in the same manner as a DPDT switch. Two special configurations are used for this task: the half bridge and the H-bridge. Let's see how they work and how can we use them in projects involving robotics and mechatronics.

4.2.1 Half Bridge

Two batteries and two SPST switches are used in the circuit of Figure 4.1. In normal operation, two conditions are possible:

1. If SW1 is closed, the current is supplied by B1, and the motor runs forward.
2. If SW2 is closed, the current flows across the motor in the opposite direction, now supplied by B2, and it runs backward.

It should be clear to the reader that *having SW1 and SW2 closed at the same time is a forbidden condition,* as it causes the two batteries run in a short circuit.

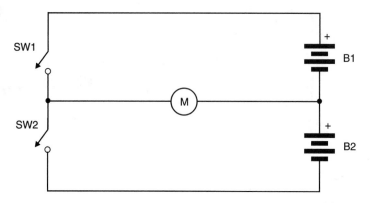

Figure 4.1 Half bridge using switches.

If we replace the switches with transistors (bipolar or FET), we can control the motor by biasing the transistors from external voltage sources as shown by Figure 4.2.

The polarity of the control voltage depends on the type of the transistor. NPN and P-channel power FETs are made conductive when positive voltages are applied to their bases or gates. Alternatively, PNP transistors are made conductive when negative voltages are applied (or they are connected to ground).

The main disadvantage of this configuration (using only two transistors) is that we need a dual or symmetric power supply. This configuration increases the degree of complexity when designing the drive circuits. The solution is given by a configuration using four transistors as shown next.

4.2.2   H-Bridge or Full Bridge

Let's start from a basic circuit where we now use four SPST switches, as shown in Figure 4.3. In operation, we can imagine two situations:

1. In Figure 4.3a, when we close SW1 and SW4, the current flows in one direction, and the motor runs forward.

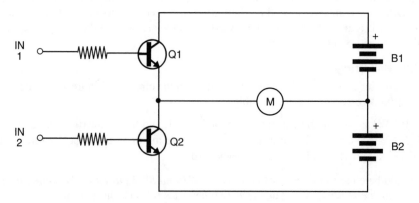

Figure 4.2 Half bridge using NPN transistors.

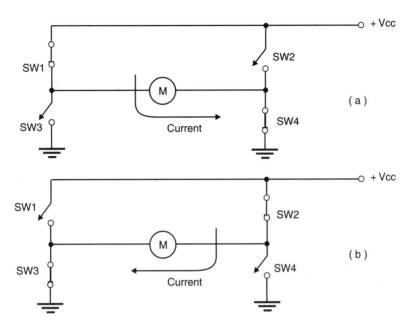

Figure 4.3   The full bridge using switches.

2. In Figure 4.3b, when we close SW2 and SW3, the current flows in the opposite direction, and the motor runs backward.

It is obvious that SW1 and SW3 can't be closed at the same time, since this procedure will cause a short in the batteries. For the same reason, SW2 and SW4 can't be closed at the same time.

Let's now replace the switches with bipolar or FET transistor as shown in Figure 4.4. Depending on the bias, we can make the transistors act as switches, conducting or not conducting the current.

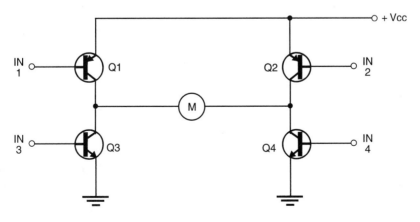

Figure 4.4   The full bridge with transistors.

In this case, we also have the forbidden condition where Q1 and Q3 conduct, making the battery run in a short, and the same condition for Q2 and Q4.

To achieve proper control of this bridge, we have two possibilities:

1. We can directly excite the transistors from sensors, switches, or another circuit in a configuration that is "intelligent" enough to avoid the forbidden conditions.
2. We can add the "intelligent" configuration or logic to the bridge, thereby avoiding the forbidden states.

In the second case, the "intelligent" configuration depends on the type of transistors used. If using PNP and NPN transistors, for instance, it is enough to wire together the bases of each transistor pair, biasing them according to the direction the motor must run. If we use only one type of transistor, NPN bipolar transistors or power MOSFET P-channel, inverters must be added to the circuit to provide the necessary logic to avoid forbidden conditions. In all these cases, we can see that the signals must be applied to the transistor so that Q1 and Q3 don't conduct at the same time (likewise, Q2 and Q4).

These two possibilities bring us to several types of circuits for dc motor control with a half bridge or full bridge, depending on the signal sources or control voltage sources.

In Figure 4.5a, we have only one input, and the motor runs in one direction when the input is high (positive voltage). It runs in the opposite direction when the input is low (ground). Notice that this configuration can be made compatible with such logic devices as TTL and CMOS.

The other option, shown in Figure 4.5b, uses two signal sources. In this case, we have four possibilities of operation, given in the following table:

IN1	IN2	Motor
High	Low	Runs forward
Low	High	Runs backward
Low	Low	Stopped
High	High	Forbidden condition

This combination of inputs can change according to the configuration—whether we use only NPN transistors, only PNP transistors, or both NPN and PNP in the same bridge. The practical blocks given in the following pages will show the reader how the configurations can change.

4.3   Practical Circuits

As in the other parts of this book, practical blocks can be used alone or combined in projects involving robotics and mechatronics. We should note that dc motors have electrical characteristics that span a very wide range of values. We find variations among models from different manufacturers, and also among units of the same type from the same manufacturer, as there may be relatively loose tolerances observed in the manufacturing process. As a result, the value of the components in many cases must be altered experimentally to suit the specific purpose of the block. In particular, the bias resistor value must be varied to obtain the correct drive of the circuit stages, based mainly on the gain of the transistors.

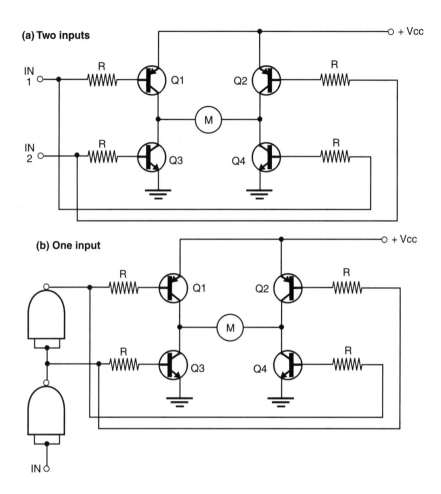

Figure 4.5   Basic configurations for full H-bridges using bipolar transistors.

Block 48   Simple Half Bridge Using Symmetric Supply

Only two bipolar transistors are necessary in a half bridge control for a dc motor if a dual or symmetric power supply is used as shown in the block of Figure 4.6. The ground is the reference in this circuit. When IN1 is positive, Q1 is saturated, and the motor is powered from B1. When IN2 is negative, Q2 conducts the power from B2, and the motor runs in the opposite direction. Notice that the reference for the signals is the ground (0 V) and two voltages (negative and positive are necessary to control the bridge). The table below shows how this circuit works:

IN1	IN2	Condition
Positive	Open or ground	Forward (B1)
Open or ground	Negative	Reverse (B2)
Open or ground	Open or ground	Stalled
Positive	Negative	Forbidden

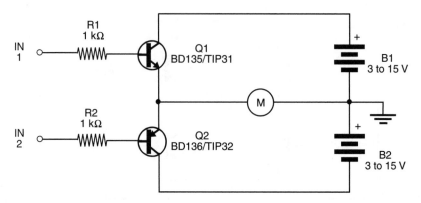

Figure 4.6   Block 48: simple half bridge.

The transistors are chosen to suit the current requirements of the motor. A table given in Section 4.5, Additional Information, shows some possibilities.

R1 and R2 depend on the transistor gain. If using BC or BD devices, these resistor can be 1 kΩ. If using TIP or other low-gain transistors, reduce these resistors to 470 Ω or even 330 Ω. Take care when driving the circuit from CMOS or TTL gates. In some cases, the TIPs need more current than these logic circuits can provide. The BD and TIP transistors must be mounted on heat sinks.

Depending on the motor, a 0.47 µF capacitor must be added in parallel. If the same batteries are used to power other circuits in the robot or mechatronics appliance, two large decoupling capacitors must be added in parallel with the batteries (1,000 µF is typical).

Block 49   Half Bridge Using Darlington Transistors

Less power is necessary to drive the circuit shown in Figure 4.7, since Darlington transistor have higher gain. This circuit needs currents in the microampere range to drive motors with currents up to a few amperes, depending on the transistors. In

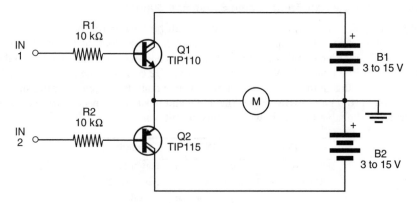

Figure 4.7   Block 49: half bridge using Darlington transistors.

Block 3, we reviewed some Darlington transistors that can be used in this application. The pair TIP110 and TIP115 can be used with motors up to 1 A. In this case, the reference for the control signals is also the ground. A 0.47 μF capacitor in parallel with the motor can be necessary in some cases. The transistors must be mounted on heat sinks. The table for control logic given in the last block is valid for this block.

Block 50   Full Bridge Using Complementary Bipolar Transistors

This circuit is the real or full H-bridge and uses four bipolar transistors to make a dc motor run forward and backward. The complete block is shown in Figure 4.8.

The sensitivity depends on the transistors used in the circuit. For the pair BD135/136, motors up to 1 A can be controlled. The gain of the transistor makes control possible from currents sources of about 5 mA or less.

Using the TIP31 and TIP32, since the gain is lower, you'll need more current to control a 1 A motor. In some cases, the resistor value must be reduced to 470 Ω. We suggest that the reader experiment with the resistor to find the best value for the application. The indicated values (1 kΩ for R1 and R2) are suitable when the signal source is a TTL or CMOS logic circuit.

The following table gives the control logic for this block:

IN1	IN2	Condition
Positive (HI)	Ground (LO)	Forward
Ground (LO)	Positive (HI)	Backward
Positive (HI)	Positive (HI)	Stalled*
ground (LO)	Ground (LO)	Stalled*

*When Q1/Q2 or Q3/Q4 conducts, they short out the motor. If the motor is running, this causes it to stop quickly.

The tables given in the previous blocks are valid when using the NPN and PNP transistor, according to the current drained by the motor. Depending on the motor, a capacitor also is needed in parallel, and the transistor must be altered to achieve the correct operation.

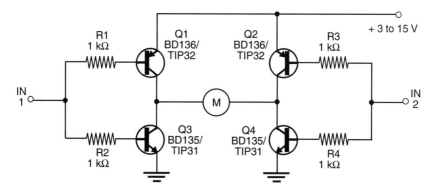

Figure 4.8   Block 50: full bridge using complementary bipolar transistors.

Block 51   Full H-Bridge with Feedback

The circuit shown by the block in Figure 4.9 uses a feedback circuit to avoid the forbidden states. When Q1 conducts, the voltage across R1 becomes positive, and transistor Q4 is forced to a saturation state. Q1 and Q4 conduct, and the motor runs forward. When Q2 conducts, the voltage across R2 becomes positive, biasing Q3 to the saturation state, and the motor runs backward.

The transistors can be the same as indicated in the previous blocks, according to the current drained by the motor. The following table shows the control logic for this circuit:

IN1	IN2	Condition
Ground	Positive	Forward
Positive	Ground	Backward
Ground	Ground	Forbidden
Positive	Positive	Stalled

The transistors depend on the current drained by the motor. See, in Section 4.5, Additional Information, a table showing common types for this task. Remember that power transistors must be mounted on heat sinks.

Block 52   Full Bridge Using NPN Transistors

The circuit shown in Figure 4.10 uses only NPN transistors and can be controlled from two logic signals. The power supply is single, and motors in the range of 1.5 to 12 V can be controlled. The transistors are chosen according to the motor. Use the table in Section 4.5, Additional Information, as reference. Remember that power transistors most be mounted on heat sinks.

The circuit is controlled by logic levels or voltages applied to the input. Be careful to avoid the forbidden condition, which can short the battery through the transistors and thus burn them out.

When using low-power transistors (BC and BD), the circuit can be controlled directly from TTL or CMOS outputs. If powerful motors are used (1 to 2 A), the circuit needs an additional buffer.

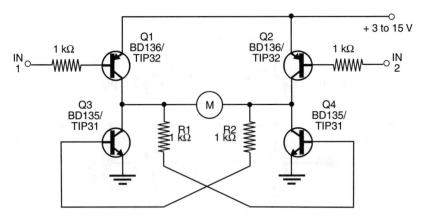

Figure 4.9   Block 51: H-bridge with feedback.

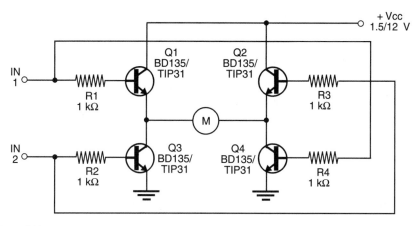

Figure 4.10   Block 52: full bridge using NPN transistors.

The following table shows the control logic for this circuit:

IN1	IN2	Condition
Positive (HI)	Ground (LO) or open	Forward
Ground (LO) or open	Positive (HI)	Backward
Positive (HI)	Positive (HI)	Forbidden
Ground (LO) or open	Ground (LO) or open	Stalled

Block 53   Full Bridge Using NPN Darlington Transistors

If high power from low power control voltages is needed, you need to use a high-gain transistor in the bridge. The circuit shown by Figure 4.11 allows the control of high power (up to 2 A or more, depending on the transistor) from low power sources such as TTL or CMOS outputs.

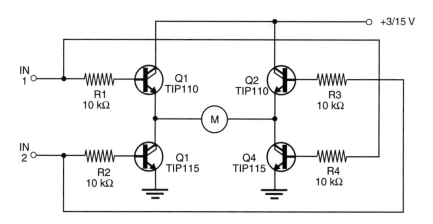

Figure 4.11   Block 53: full bridge using NPN Darlington transistors.

The Darlington transistors depend on the motor. The tables given in Section 4.5, Additional Information, can help the reader to choose the ideal transistor for the desired application.

The control logic is the same as in the previous block. The basic difference is that, due the gain of the transistor, you will need only few milliamperes, or even microamperes, to drive the circuit when in the high or positive logic level.

Depending on the transistors and the power of the motor (current drain), the resistor must be varied in the range between 2.2 kΩ and 22 kΩ. Conduct experiments to achieve the best performance.

Block 54   Full Bridge with Logic Control

As we have seen, the direction of the current across the motor in an H-bridge depends on two input signals: one must be positive and the other negative simultaneously. To provide the necessary inversion function for one input, a logic circuit can be added to an H-bridge as shown in Figure 4.12.

The circuit uses two of the four gates of a CMOS 4093 IC. Other logic functions can be used, including TTL. (See Section 4.5, Additional Information, for some integrated circuits that are useful for this task.)

When the input signal is a high logic level (positive), the output of the first gate is low, and Q2 conducts. At the same time, the output of the second block is high, and Q3 conducts. The current can flow from Q2 to Q3, making the motor run forward. If the input passes to the low logic level, Q1 and Q4 conduct, and the current flows from Q1 to Q4, making the motor run backward. Note that the forbidden condition

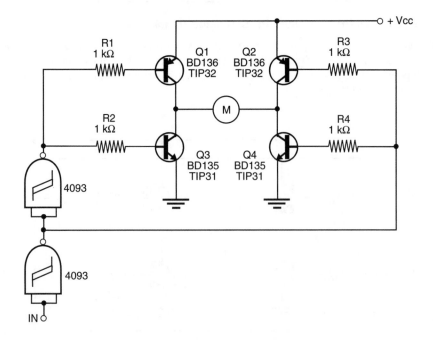

Figure 4.12   Block 54: full bridge with logic control.

is avoided, but the transistor can only run forward and backward. The stalled condition doesn't exist and may be implemented by an additional block.

The recommended transistors for motors up to 1 A are the BD135 mounted on heat sinks. See Section 4.5, Additional Information, for other transistors and concerns with this circuit.

Block 55   Full Bridge with Logic Control and External Enable

In this block, we find the additional resource of an external enable. The enable input is IN2. Setting this input at the high logic level or applying a positive voltage to it, the circuit is enabled, and the motor runs forward or backward, depending on the logic level applied to IN1. The circuit is shown in Figure 4.13, and the transistor can be chosen using the tables in Section 4.5, Additional Information, as in previous blocks.

Q5 must be the same as Q3 and Q4, depending on the current in the motor. Notice that we must consider the voltage drop in three transistors when the motor is on. This voltage drop must be compensated by an increase in the power supply voltage. This compensation is especially important when working with low-voltage motors (below 9 V).

Block 56   Power MOSFET H-Bridge

The low Rds(on)[*] characteristic of power MOSFETs makes these devices especially appropriate for power control, as in the case of H-bridges. The circuit shown by Figure 4.14 is the basic configuration using N-channel power MOSFETs.

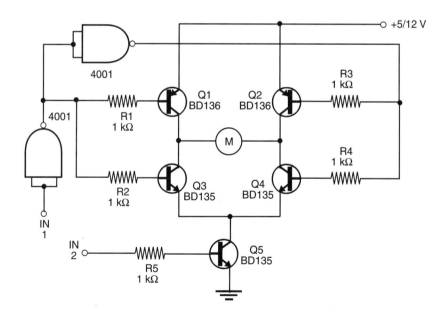

Figure 4.13   Block 55: full bridge with enable input.

---

[*] Resistance between drain and source pins when turned on.

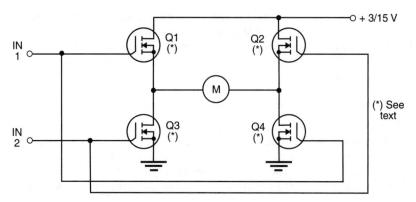

Figure 4.14   Block 56: full bridge with power MOSFETs.

Power MOSFETs with Rds(on) below 1 Ω are common. These transistors can control currents in the range between 1 A and 20 A in circuits like this one.

The transistors for a specific application are selected according to the current in the motor. They must be mounted on heat sink, especially if you're controlling high-power motors (above 2 A).

This circuit operates in exactly the same way as Block 62. The following table gives the logic of the control.

IN1	IN2	Condition
High (positive)	Low (ground or open)	Forward
Low (ground or open)	High (positive)	Reward
High (positive	High (positive)	Forbidden
Low (ground) or open	Low (ground) or open	Stalled

Be careful to avoid the forbidden condition, which shorts out the power supply. Using a fuse in series is a good way to keep your circuit from becoming a smoke factory.

Block 57   H-Bridge Using Power MOSFETs with Enable

Adding a power MOSFET as shown in Figure 4.15, the H-bridge using power MOS-FET can be enabled by an external logic signal. Q5 controls all the current across the bridge and is on when the input IN3 is low, and off when the input is high. The control logic is given by the following table:

IN1	IN2	Condition
High	Low	Forward
Low	Low	Reward
Don't care	High	Stalled

Q5 is the same transistor as all others in the bridge, and it is chosen according to the current drained by the motor.

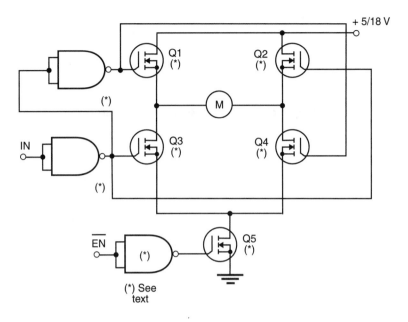

Figure 4.15   Block 57: full bridge with enable input.

Block 58   Combined Bipolar + MOSFET H-Bridge

The configuration shown by Figure 4.16 combines bipolar transistors with power MOSFETs. R1 and R2 form a feedback network that determines which bipolar transistor conducts when the power MOSFETs are triggered on. The value of these re-

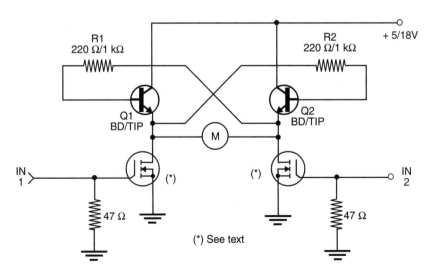

Figure 4.16   Block 58: combined bipolar + power MOSFET.

sistors depends on their gain. Using high-gain transistors such as the BD136, BD138, or another, you can use a 1 kΩ resistor. For transistors with less gain, e.g., the TIP32 or TIP42, you'll need 220 Ω resistors.

All the transistors need heat sinks. The circuit has the same control logic as one shown for Block 52.

Block 59   Combined Darlington + Power MOSFET H-Bridge

The circuit shown in Figure 4.17 combines MOSFETs with bipolar Darlington transistors. This block operates in the same manner as described for the previous block. The feedback resistors can have higher values in this case. The transistors can be chosen using the tables given in Section 4.5, Additional Information.

Block 60   R-S Flip-Flop H-Bridge

The configuration shown by Figure 4.18 is an H-bridge that is controlled by pushbuttons or sensors, operating as a flip-flop. Touching SW1 makes the motor run forward. Momentarily pressing SW2 makes the motor runs backward. To reverse the motor's direction again, it is enough to press SW1 momentarily.

R1 and R2 are chosen according to the gain of Q1 and Q2. See Block 58 for more details.

The bipolar transistor and the power MOSFETs are chosen according to the current drained by the motor. If heavy-duty motors must be controlled, the transistors must be mounted on heat sinks. The circuit can be powered from supplies in the range of 9 to 15 V.

Block 61   Complete H-Bridge

This block (Figure 4.19) combines power MOSFETs with common bipolar NPN transistors and operates with two inputs. A feedback circuit provides protection against the forbidden states. When this state is forced by the inputs, the feedback to the base of Q3 and Q4 does not allow them to be conductive at the same time.

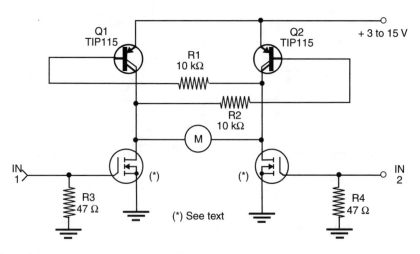

Figure 4.17   Block 59: Darlington + MOSFET H-bridge.

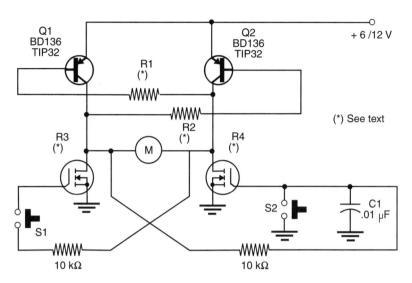

Figure 4.18   Block 60: flip-flop bridge.

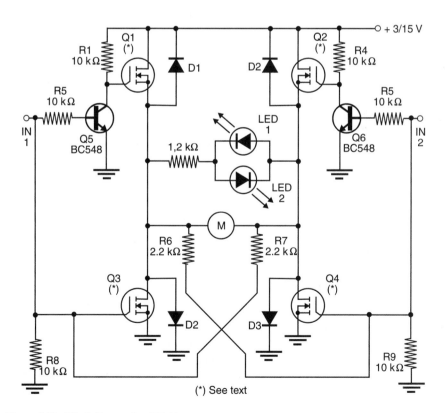

Figure 4.19   Block 61: complete H-bridge.

The Power MOSFETs can be chosen according to the motor using the tables in Section 4.5, Additional Information. The bipolar transistors are BC548 or any other general-purpose NPN silicon transistor.

The two LEDs act as direction indicators. The value of resistor R3 depends on the voltage used to power the circuit according to the following table:

Vcc	R1
6 V	470 Ω
9 V	680 Ω
12 V	1 kΩ
15 V	1.8 kΩ
18 V	2.2 kΩ
24 V	2.7 kΩ

The control voltage can be supplied by common TTL or CMOS gates or operational amplifiers.

4.4   Integrated H-Bridges

The circuits shown in the previous blocks used discrete components (i.e., transistors, resistors, diodes, and other components). This can be considered an important solution in didactic works. But it isn't the only solution for projects in this area.

The modern designer of robotics and mechatronics circuits can find many H-bridges in the form of ICs. Many manufacturers, in their line of ICs, offer complete H-bridge circuits that can control high-power motors, and they can be used with no (or few) additional external components.

Two of the most popular IC H-bridges are listed below, and afterward we offer some application blocks.

Type	Characteristics	Manufacturer
LMD18200	3 A, 55 V	National Semiconductor
LMD18201	3 A, 55 V	National Semiconductor

Internet search engines can help the reader find more information about these and other ICs.

Block 62   H-Bridge Using the LMD18200

The LMD18200, manufactured by National Semiconductor, is a 3 A, 55 V bridge designed for motion control applications. This IC uses a multi-technology process combining bipolar and CMOS control circuits with DMOS power devices on the same chip. This circuit is useful to drive both dc motors and stepper motors.

In the block shown by Figure 4.20, we have an application suggested by National Semiconductor in which the LM18200 is used to control the current through the motor by applying an average voltage equal to zero to the motor terminals for a fixed period of time, whenever the current through the motor exceeds the commanded current. This action causes the motor current to vary slightly about an externally controlled average level. The duration of the off-period is determined by the RC net-

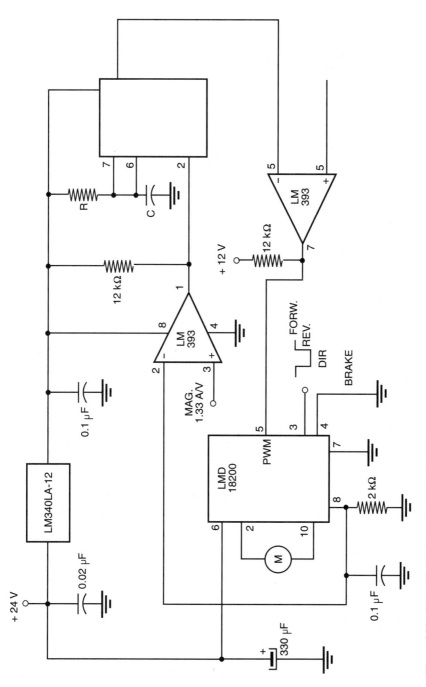

Figure 4.20 Block 62: H-bridge using the LMD18200.

work in the 555. In this circuit, the sign/magnitude mode of operation is implemented [see, in the next chapter, how pulse width modulation (PWM) controls
operate]. The switching waveforms in this circuit are shown by Figure 4.21.

Block 63   H-Bridge Using the LM18201

National Semiconductor's LM18201 is a 3 A, 55 V H-bridge in a single chip, combining bipolar and CMOS devices in the control section and power DMOS transistors in the output. The block shown in Figure 4.22 includes a current-sensing output
that can be used to drive tachometers or speed regulators.

The PWM input (see next part) is used to control the speed of the motor. The input brake gives the motor a fast stop by shortening its terminals. The PWM signals
can be sign/magnitude or simple, locked antiphase PWM.

In the first type (sign-magnitude), the control pulses consists of separate direction
(sign) and amplitude (magnitude) signals. The absolute magnitude signal is duty-cycle modulated, and the absence of a pulse signal (logic level low) represents zero
drive.

In the second type (simple, locked antiphase PWM), the control pulse consists of
a single variable duty-cycle signal in which both direction and amplitude information is encoded. A 50% duty cycle PWM signal represents zero drive, since the net
value of voltage delivered to the load is zero.

4.5   Additional Information

The transistors to be used in a half or full bridge depend on the current that they
must control. This current is determined by the motor and can vary from few milliamperes to more than 2 A. In the following tables, we present the reader with some
transistors that are suitable for applications in bridges.

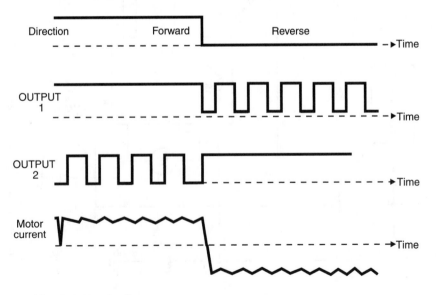

Figure 4.21   Switching waveforms.

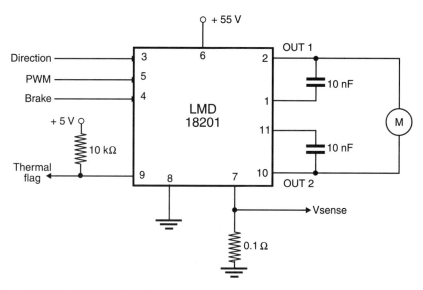

Figure 4.22   Block 63: H-bridge with the LM18201.

NPN Bipolar Transistors

Type	Voltage (Vce)	Current (Ic)	Gain (hFE)
BC547/548/549	45/30/30 V	100 mA	110-800/110-800/200-800
2N3903/3904	40 V	200 mA	20-200
BD135/137/139	40/60/80 V	1.5 A	40-250
BD433/435/437	22/32/45	4 A	85-475/85-475/85-375
TIP31/A/B/C	40/60/80/100 V	3 A	10-50
TIP41/A/B/C	40/60/80/100 V	6 A	15-75

PNP Bipolar Transistors

Type	Voltage (Vce)	Current (Ic)	Gain (hFE)
BC557/558-559	45/30/30 V	100 mA	75-800/75-800/125-800
2N3905/3906	40 V	200 mA	20-200
BD136/138/140	40/50/80 V	1.5 A	40-250
BD434/436/438	22/32/45 V	4 A	85-475/85-475/85-375
TIP32/A/B/C	40/60/80/100 V	3 A	10-50
TIP42/A/B/C	40/60/80/100 V	6 A	15-75

NPN Darlington Transistors

Type	Voltage (Vce)	Current (Ic)	Gain (hFE)
BD331/333/335	60/80/100 V	6 A	750
TIP110/111/112	60/80/100	1.25 A	500
TIP120/121/122	60/80/100	5 A	1,000
TIP140/141/142	60/80/100	10 A	1,000

PNP Darlington Transistors

Type	Voltage (Vce)	Current (Ic)	Gain (hFE)
BD332/333/335	60/80/100 V	6 A	750
TIP115/116/117	60/80/100 V	1.25 A	500
TIP125/126/127	60/80/100 V	5 A	1,000
TIP145/146/147	60/80/100 V	10 A	1,000

Power MOSFETs (N-channel)

Type	Voltage (Vds)	Current (Id)	Rds(on)
IRF222	200 V	4.0 A	1.2 Ω
IRF230	200 V	9.0 A	0.4 Ω
IRF250	200 V	30 A	0.085 Ω
IRF223	150 V	4.0 A	1.2 Ω
IRF231	150 V	9.0 A	0.4 Ω
IRF122	100 V	7.0 A	0.4 Ω
IRF142	100 V	24 A	0.11 Ω
IRF123	60 V	7.0 A	0.4 Ω
IRF143	60 V	24 A	0.11 Ω
IRF151	50 V	40 A	0.055 Ω

4.6   Special Recommendations

4.6.1   Decoupling Capacitors

When turning an inductive load (e.g., a dc motor) on and off, high-voltage and high-current spikes are generated, propagating across the circuit and potentially causing problems. To avoid damaging spikes, robotics and mechatronics designers must exercise much care.

Decoupling capacitors must be included at all points where current or voltage spikes can be present and threaten the stability of the circuit. Some points of the circuit are especially important when looking for a place to add a decoupling capacitor.

In parallel with the power supply.   Large electrolytic capacitors must be added in parallel with the power supply as shown by Figure 4.23. These capacitors store energy and add it to the current supplied by the source when the motor is turned on and draws high currents. Values between 1,000 and 10,000 µF are suitable for common applications. If the motor draws very high currents on startup, and the voltage drop in the power supply can't be avoided, it is recommended that you use a separate battery to power it.

In parallel with the motor.   Capacitors in the range between 0.1 and 1 µF (polyester types) are recommended to absorb the spikes generated by the commutation system of the motor. These capacitors will protect the switching devices against high-voltage spikes produced when the motor is running.

4.6.2 Transistor Protection

Transistors are very sensitive to voltage spikes such as the ones generated by inductive loads. Many types of transistors used in switching include internal protective diodes. But if they do not, it is recommended that you add them as shown in Figure 4.24.

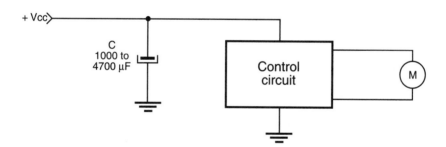

Figure 4.23   Decoupling capacitor.

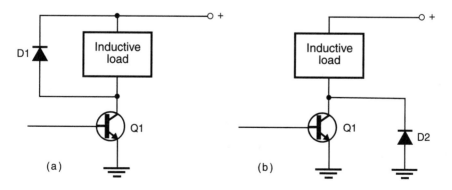

Figure 4.24   Protecting the transistor.

The ideal components for this purpose are Schottky diodes, since these devices are fast enough to react to the short pulses generated by the dc motors. But, for common applications, your circuits can be protected by common diodes such as the 1N4002 and others in that series.

4.6.3 Current Sensing

In some applications, it is important to add some kind of circuit to monitor the current drained by a motor. This circuit can be used to control the speed or to manage the use of power, extending the battery life. Figure 4.25 shows how this circuit can be added.

A resistor of 0.1 Ω causes a voltage drop of 100 mV for each ampere drained by a motor. Adding an operational amplifier with gain 10, for instance, you'll have 1 V/A directly driving some control circuit or monitor circuit. The value of this resistor must be as low as possible so we don't "steal" energy from the motor.

4.6.4 Fuses

Many blocks shown in this chapter have "forbidden states" that can cause the power supply to short out. The high current passing across the transistors in these conditions can burn them.

To protect the circuit against these conditions, fuses are recommended. The fuses are placed in series with the power supply, and their values depend on the motors and the transistors. It is a good rule of thumb to use fuses 2.5 to 3 times greater than the normal current drained by the motors.

4.7   Suggested Projects

As in all other parts of this book, many blocks can be tied together in robotics and mechatronics projects. Some suggestions for combined blocks follow:

- The delayed or timed relay given in Blocks 39 and 40 can be combined with any H-bridge. The motors will be turned on and off or reversed after a certain period of time.

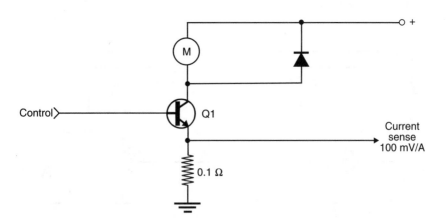

Figure 4.25   Current sensing.

- Blocks 11 and 12 can be used to add speed to motors that are controlled by an H-bridge. (In the next block, we will see how to control the speed using transistors.)
- Block 15 or 24 can be used with any H-bridge to change not only the direction of one motor but also of two motors, selected by an external control.
- Blocks between 30 and 37 can be used in series with an H-bridge as an enable control.

4.8   Review Questions

1. How many transistors should become conductive when an H-bridge is on?
2. How many transistor must be used in a half bridge?
3. In the forbidden condition of an H-bridge, what happens with the current flowing from the power supply?
4. What is necessary for the logic at the input of an H-bridge?
5. How many operational states does a two-input H-bridge have?
6. Is it possible to use power MOSFET and bipolar transistors in the same bridge? Why or why not?

5

Linear and PWM Power Controls

5.1 Purpose

In previous chapters, we have seen how to change the direction of a motor using various configurations according to the needs of the designer. But reversing the direction of a motor is not the only type of control we need for projects in robotics or mechatronics. Speed control of the motor is as or more important in many projects. In this chapter, we provide the reader with some techniques that can be used to control the speed of a dc motor. The same principles can be applied to the control of other loads such as the force of a magnet or solenoid, the temperature of a heater, or the brightness of an incandescent lamp.

5.2 Theory

The simplest way to control the power applied to a load is via the use of a rheostat wired in series as shown by Figure 5.1. The rheostat and the load form a variable voltage divider. By changing the resistance of the rheostat, the voltage applied to the motor also changes, and consequently so does its speed.

This type of control, although very simple, presents some inconveniences:

1. The same current drained by the motor passes across the rheostat. Depending on the size and power of the motor, this means that a large amount of heat is generated in the rheostat. Special heat dissipation methods must be employed, increasing the size and the cost of the component.
2. When controlling a dc motor, since the motor represents also a variable load (the current across it depends on the load and the speed), the circuit is unstable; it is

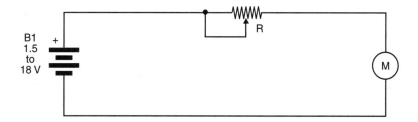

Figure 5.1 Simplest motor control.

not easy to make the motor start softly, keeping the torque constant. The tendency is for a hard start, making the robot or other mechanism jerk forward.

These two inconveniences can be eliminated by using special configurations that will be described in this chapter. As in other chapters, the blocks are basic and must be adjusted so that their characteristics suit the particular applications and motors. All the data needed to accomplish this task will be given in the following pages.

5.2.1   Two Types of Controls

The two circuit configurations described below can be used to control the power applied to a motor or other load using modern components to avoid the previously described problems. (A third kind of control is described in the next chapter. It uses special components that are described in detail therein.)

5.2.2   Linear Controls

The linear control uses a transistor (bipolar or FET) as a variable resistor in a typical configuration such as shown in Figure 5.2. The base current changes the resistance between collector and emitter and thereby the current flowing across the circuit.

The main advantage of this circuit is that the base current is low as compared to the current flowing across the transistor, meaning that this component doesn't need to dissipate much power. The transistor is the element that dissipates the power. This allows the use of common carbon, low-dissipation potentiometers to control high-current loads.

The disadvantage is the one discussed before: we experience power loss, as the transistor converts a great deal of power into heat while controlling the current across the load. The power converted into heat is given by the voltage drop across the transistor times the current controlled, as shown in Figure 5.3.

For instance, if a transistor is adjusted to apply 6 V to a 0.5 A motor from a 12 V power supply as shown by the figure, the voltage drop of 6 V across the transistor times the current (0.5 A) will result in the production of 3 W of heat.

5.2.3   Pulse Width Modulation

Pulse width modulation (PWM) is a much more efficient power control technology. Therefore, it is used not only in the control of dc loads but also in many other applications (e.g., as power supplies).

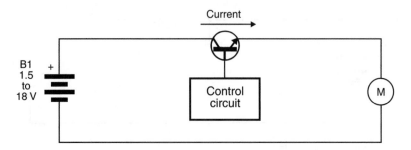

Figure 5.2   Linear control using a bipolar transistor.

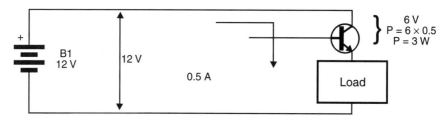

Figure 5.3   Large amounts of power are converted into heat by the transistor.

The basic idea starts from the use of square voltage pulses to power a load as shown by Figure 5.4. The amount of power applied to the load depends of the duration of each pulse or the *duty cycle* of the signal.

If the duration of the pulse is the same as the interval between pulses (representing a duty cycle of 50%, as shown in Figure 5.4a), the average power applied to the load is 50%. If we extend the duration of the pulse, the average power applied to the load increases by the same proportion, as shown in Figure 5.4b.

By controlling the *width* of the pulses, we can control the power applied to a load. The process used to control the width of the pulse is called *modulation,* and this kind of circuit termed a *PWM* or *pulse width modulation* power control.

How a Practical PWM Control Circuit Works.   Let's start with the configuration shown by Figure 5.5. A power transistor (MOSFET or bipolar) is wired to the output of a variable-duty-cycle oscillator. When the oscillator is running, the transistor turns on and off at the same frequency, applying a squarewave voltage to the load. The average voltage in the load, as we have seen, depends on the duration of the pulses. The great advantage to this type of circuit is that the power dissipated by the transistor is near zero.

When the transistor is on, its resistance can be considered to be zero, and no power is generated across it. (The power is the product of the voltage drop and the current, and since the voltage fall is practically zero, the result is zero power.) On the

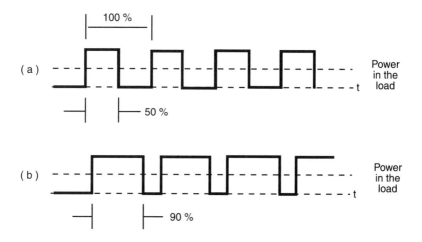

Figure 5.4   The power depends on the pulse width.

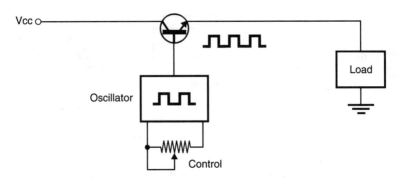

Figure 5.5   The basic PWM control.

other hand, when the transistor is off, no current flows through it, and again the power dissipated is zero.

When working with "real world" components, things are quite different. The transistor can't pass from the on state to the off state, or vice versa, quickly enough to avoid dissipation problems. The transistor needs a finite time for this state change.

As shown in Figure 5.5, during the time the transistor switches from one state to another, the resistance presented by the device changes, and current and voltage drops simultaneously appear in the circuit. These are large enough to cause some heat to be generated.

The heat generated by the switching process can be important in this kind of circuit, but it is very low as compared to the heat generated by linear controls. Another important advantage of PWM circuits when controlling dc motors is that they maintain more constant torque over the entire speed range.

As we have seen, when using a linear control, the dc motor tends to jerk forward when it draws enough power to overcome inertia. Using a PWM control, the pulses always contain the total circuit voltage, and only the pulse duration changes. This means that, even at very low speeds, the motor receives a high enough voltage to overcome inertia and begins to rotate. It is therefore possible to achieve "soft" adjustments throughout the speed range, which is not possible with a linear control.

5.2.4   Two Forms of PWM Controls

Two forms of PWM controls can be found in practical applications involving robotics and mechatronics. They are described below.

Locked anti-phase control.   The simple locked anti-phase control consists of a single, variable-duty-cycle signal oscillator in which both direction and amplitude (speed) information is encoded, as shown in Figure 5.6. A 50% duty-cycle signal represents zero drive, since the net value of voltage (integrated over one period) delivered to the load (motor) is zero.

The great disadvantage of this kind of control is that, when the applied power is zero, the power supplies are dispensing power 50% of the time. This power is converted into heat. For this reason, this kind of control is not recommended for controlling high-power motors.

Sign/magnitude control.   This control employs separate direction (sign) and amplitude (magnitude) signals. The magnitude signal is duty-cycle modulated, and the absence of a pulse signal (a high logic level) represents zero drive. Figure 5.7 shows

Keep up-to-date with the latest books in your field!

Complete this postage-paid reply card and return it to us now! We will notify you of upcoming titles and special offers. Visit our website and check out information on our newest releases.

What title have you purchased? _____

_____

How did you hear about it? _____

Where was the purchase made? _____

Name _____

Job Title _____

Institution _____

Address _____

City _____ State _____

Zip/Postcode _____

Country _____

Telephone _____

Email _____

❏ Please arrange for me to be kept informed of other books and information services on this and related subjects. This information is being collected on behalf of Reed Elsevier Inc. and may be used to supply information about products by companies within the group.

(FOR OFFICE USE ONLY)

www.bh.com

BRC2001

BUSINESS REPLY MAIL

FIRST-CLASS MAIL   PERMIT NO. 78   WOBURN MA

POSTAGE WILL BE PAID BY ADDRESSEE

DIRECT MAIL DEPARTMENT
Butterworth-Heinemann
225 WILDWOOD AVE
PO BOX 4500
WOBURN MA 01888-9930

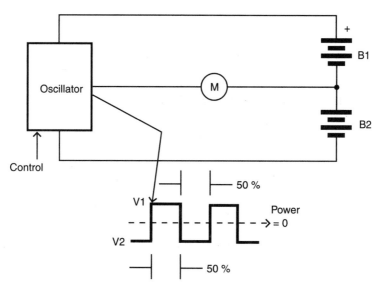

Figure 5.6   Locked anti-phase control.

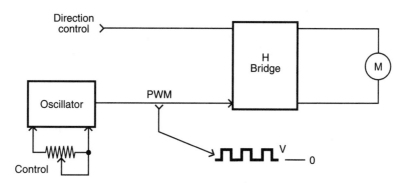

Figure 5.7   The sign/magnitude PWM control.

what happens with this control. Current delivered to the load is proportional to the pulse width.

5.3   Basic Blocks

5.3.1   Linear Controls

Linear and PWM power controls can be designed with many common components. The following blocks are examples of configurations that can be used in projects involving robotics, mechatronics, and other forms of automation.

All of these blocks use common parts and can be adapted to specific tasks. The components in most cases can be altered to match the circuit with the characteristics of the loads to be controlled. We invite the reader to experiment with these blocks, substituting components to achieve the best performance in each case. Suggestions about the range of values appropriate to the critical components are given with each block.

Block 64   Electronic Rheostat

The block shown in Figure 5.8 can be used to control the power applied to a dc motor, an incandescent lamp, or any other inductive or resistive load. The transistor is used as a variable resistor that depends on the bias given by the potentiometer. The advantage to the use of the transistor is the low current in the potentiometer. Common carbon potentiometers can be used to control heavy-duty loads up to several amperes, depending on the load.

The transistor must be mounted on heat sinks. The tables of transistors in Chapter 4, Section 4.5, can be used as reference to choose the proper transistor. For motors up to 1 A, we recommend the BD135 or TIP31. For large motors, the 2N3055 can be used.

The resistor value can be altered to give total power when the potentiometer is completely open. But it is not recommended that you reduce the value of this resistor to below 220 Ω. This resistor must have a dissipation at least of 1 W if the 2N3055 is used and the power supply voltage is 12 V or more.

Loads powered from voltages in the range between 3 and 18 V can be controlled with this circuit. Remember that this is a linear control, and the torque will not be linear along the voltage range applied to the load, as discussed in Section 5.2.

If inductive loads such as dc motors are controlled, a diode must be wired in parallel to protect the transistor against voltage spikes generated when the load is powered up.

Block 65   Linear Control Using a Darlington Transistor

A linear control design can be simplified by using a Darlington transistor as shown in Figure 5.9. As discussed earlier in this book, Darlington transistors are formed by two transistors in a direct coupled stage inside a unique package. Darlington transistors that can control current up to several amperes are cheap and easy to obtain from any component dealer.

The tables of transistors in Chapter 4, Section 4.5, can be used to choose the Darlington NPN transistor suitable for your application. Remember that the transistor must be mounted on a heat sink.

The value of resistor R1 must be established experimentally to match the gain of the transistor with the desired range of control. Values in the range between 4.7 and 1 MΩ are good candidates.

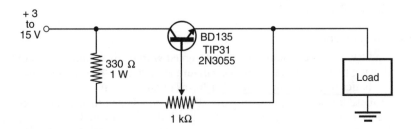

Figure 5.8   Block 64: rheostat.

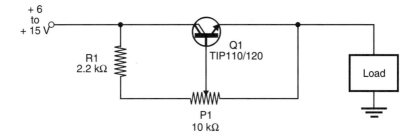

Figure 5.9   Block 65: linear control using a Darlington transistor.

Block 66   Linear Power Control Using a Zener Diode

As we have seen, to control the speed of a motor, the brightness of an incandescent lamp, or the power applied to a load, we can use a linear control circuit that is nothing more than a variable voltage source or power supply. Many configurations exist for variable power supplies using discrete parts or integrated circuits. The designer will find them useful in applications involving power control as the one need in robotics and mechatronics works

The variable power supply shown in Figure 5.10 is a traditional configuration using a zener diode as voltage reference. This block can also be used in applications such as a workbench power supply used to test circuits and devices that require currents up to 1 A. You can add a rectifier stage and a transformer as shown in Figure 5.11 to convert it into a complete power supply.

The presence of the zener diode is important in defining the voltage range supplied by the circuit. This range starts at 0 V and has upper limit defined by the zener diode voltage plus 0.6 V (the voltage drop in the emitter-base junction of the transistor). For instance, if you use a 12 V zener diode, your supply will provide voltages in the range between 0 V and 12.6 V.

Using a 6 V × 400 mW diode, you can control the speed of 6 V loads without the risk of burning out the motor with excess voltage, even if the input is much more than 6 V. The recommended input voltage for this configuration is at least 2 V higher than the nominal voltage in the output and twice this voltage. For instance, if the load is a 6 Vdc motor, the recommended input for this application is between 8 and 12 V.

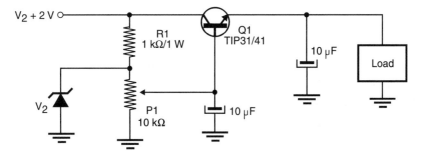

Figure 5.10   Block 66: linear control using a zener diode as reference.

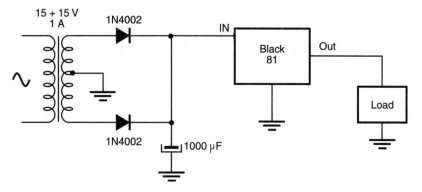

Figure 5.11   Converting Block 66 into a power supply.

The transistor is chosen using the table in Chapter 4, Section 4.5, and any 400 mW or 1 W diode can be used for output currents up to 1 A. The transistor must be mounted on heat sink.

Zeners in the range between 3 V and 12 V can be used. Resistor R1 can be experimentally valued in the range between 470 Ω and 2.2 kΩ to allow the circuit to cover the desired voltage range.

5.3.2   Constant Current Sources

An important group of linear power control blocks are the constant current sources. These are circuits that can maintain a constant current across a load even when its resistance is changed from external causes. This is the case with motors whose resistance drops when their load increases, which produces a tendency to drain more current from the voltage source. In some cases, limiting this current can be important to avoid overloading the circuit.

Another case is found when shape memory alloys (SMAs) are powered, and the current across them must be kept to within a narrow band of values even when the control voltage changes.

Finally, we can consider cases in which incandescent lamps need to maintain constant brightness even when the input voltage changes, and heaters whose resistance changes with temperature (resistance drops when the temperature falls) and thus require some kind of regulation in to be suitable for robotics, automation, or mechatronics applications.

The constant current source can be used to maintain a predetermined current value in a load even when the resistance of the load is altered or when the input voltage of the circuit is changed. The following blocks can be used for this task.*

Block 67   Constant Current Source Using Transistor

An interesting block for robotics and mechatronics designers is the one shown in Figure 5.12. This circuit maintains a constant current flow across a load. This can be

---

* Remember that constant *voltage* sources here are the batteries normally used to power the robots and other appliances. These are different from the current sources. Some constant voltage sources are described in other parts of this book.

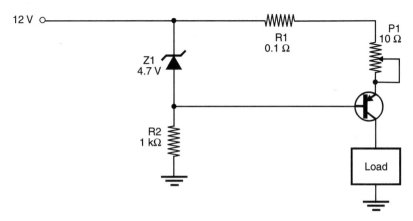

Figure 5.12   Block 67: constant current source using transistor.

important to keep the motor speed constant even when its load changes, or to keed the current across a heater constant when its temperature changes, thus altering its resistance.

The zener diode is a 400 mW or 1 W type, and P1 is used to adjust the current across the load. This circuit can be used to maintain constant currents up to 500 mA. If larger currents are involved, a Darlington transistor can be used.

P1 must be a wire-wound potentiometer, and its dissipation rate is important if larger loads are controlled. The current adjustment can be made using a current meter in series with the load.

If higher voltages are used at the input, the 1 kΩ resistor must be increased in value, and the zener must have larger dissipation power. If the load is a motor, it is convenient to add a decoupling capacitor in parallel. Electrolytic capacitors between 1 and 470 μF are recommended for this task.

Block 68   Constant Current Source Using the LM350T (3 A)

A current up to 3 A can be kept constant across a load using the block shown in Figure 5.13. The LM350T is a three-lead TO220 packaged IC consisting of a variable voltage regulator with output currents up to 3 A. Using the configuration shown in the figure, the circuit operates as a current regulator, keeping the current constant across a load.

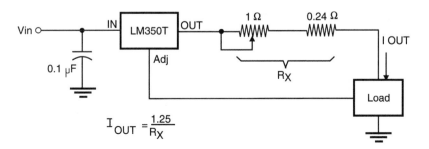

Figure 5.13   Block 68: constant current source using the LM350T (3 A).

The potentiometer is used to adjust the current across the load, and the resistor limits its highest value. The IC must be mounted on a heat sink, and the input voltage typically must be in the range between 12 and 24 V. Other ICs in the same series as the LM150T and LM250T can be used in this block.

Block 69   Variable Constant Current Source Using the LM338 (5 A)

This block employs a potentiometer to adjust the current in the load. The circuit is shown in Figure 5.14.

The LM338 can source up to 5 A to a load. The current can be calculated as function of the resistor Rx by the formula with the diagram. A large heat sink is needed for the applications in which the IC operates near the power limit.

5.3.3   PWM Blocks

The following are blocks of PWM controls using few components and suitable for motors up to several amperes. The blocks can be altered to match the characteristics of the motors and the applications, since the speed and the load can have some influence on the circuit.

Block 70   PWM Basic Control Using the 4093 CMOS IC

The block shown in Figure 5.15 is a simple PWM control for dc motors up to several amperes, depending on the transistor chosen for Q1 (see tables in Chapter 4, Section 4.5). The circuit is formed by a low-frequency oscillator using one of the four gates of a 4093 IC. The other three gates are wired as a buffer/amplifier to drive the output transistor.

Capacitor C1 determines the frequency range and must be chosen to suit the motor. In some cases, some motor vibration can occur if an improper capacitor is used.

Resistor R1 determines the smallest pulse width applied to the motor, and the ratio between P1 and R2 sets the maximum duty cycle of the control. Experiments with R1 and R2 must be conducted to find the best operating range for any motor.

D1 and D2 can be any general-purpose silicon diode such as the 1N914, 1N4148, or even 1N4002. Pin 1 of the IC can be used as an external enable input. The circuit can be used with Darlington transistor if you replace R3 with a 4.7 kΩ resistor.

Block 71   PWM Control Using the 4001/4011 CMOS IC

The block shown in Figure 5.16 is formed by a low-frequency oscillator using two of the four gates of a 4001 or 4011 CMOS IC. The output transistor (Q1) is chosen

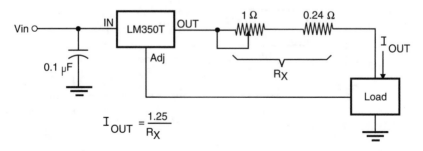

Figure 5.14   Block 69: constant current source with the LM338.

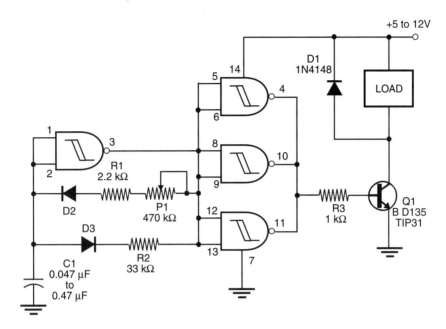

Figure 5.15   Block 70: basic PWM control.

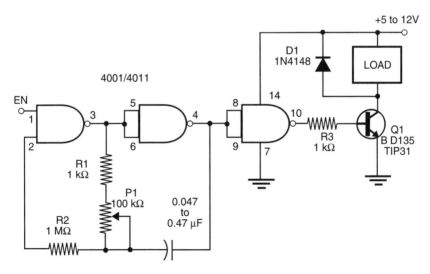

Figure 5.16   Block 71: PWM control using the 4001/4011 CMOS IC.

to match the characteristics of the dc motor to be controlled. (See the tables in Chapter 4, Section 4.5.) If a Darlington transistor is used, R3 can be increased to 4.7 kΩ.

P1 determines the power range. R1 and R2 must be chosen to give the desired power range. C1 also depends on the motor and is chosen to avoid vibrations during motor operation.

The other two gates of a 4011 or 4001 IC can be used in other applications, since they are functionally independent. This circuit can also be altered to use a power MOSFET to replace Q1. The gate of the transistor is then plugged directly to the output of the three gates of the IC.

Block 72   Medium/High-Power PWM Control Using the 555 IC

Adding a bipolar power transistor, we can control loads up to several amperes as shown by the block in Figure 5.17. The NPN transistor is chosen to suit the motor requirements. See the tables in Chapter 4, Section 4.5, to find the correct transistor for your motor.

As in previous blocks, the value of resistors R1 and R2 must be found experimentally depending on the characteristics of the motor. The same is valid for C1.

Darlington NPN power transistors can be used in this block. In that case, increase the value of R3, replacing it with a 4.7 kΩ or even a 10 kΩ resistor. The transistor must be installed on a heat sink.

Block 73   Medium-Power PWM Using the 555 IC and a PNP Transistor

In the previous block, the motor is on when the output of the IC is at the high logic level. We can invert this operational principle using a PNP transistor such as the one shown in Figure 5.18.

The transistor is chosen using the tables in Chapter 4, Section 4.5, according to the current requirements of the motor. The same components must be valued experimentally to achieve a correct match with the motor's characteristics. Darlington PNP transistors can be used in this circuit; replace R3 with a 4.7 or 10 kΩ resistor.

Block 74   Anti-phase PWM Power Control Using the 555 IC

The circuit shown in Figure 5.19 can be used to control both the speed and direction of a small dc motor. Motors with currents up to 100 mA can be controlled by this

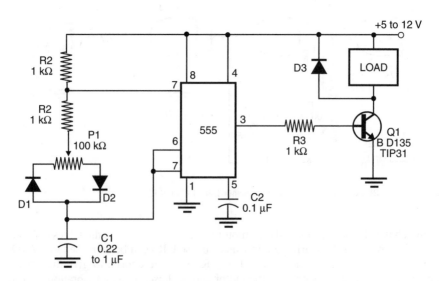

Figure 5.17   Block 72: medium/high-power PWM control.

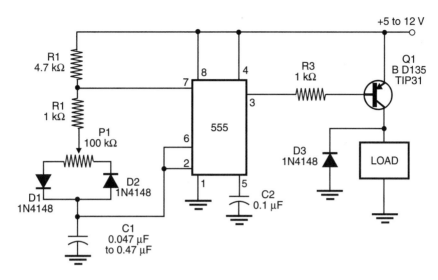

Figure 5.18   Block 73: medium-power PWM using a PNP transistor.

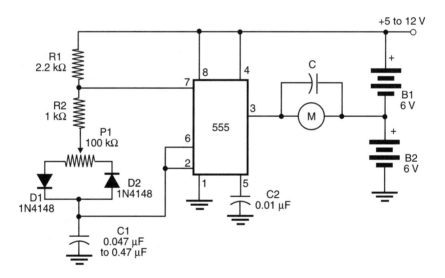

Figure 5.19   Block 74: anti-phase PWM, low power.

circuit. This circuit needs a symmetric power supply or a 6 + 6 V power supply as explained previously.

When the output of the 555 is at the high logic level, B1 supplies the current to the load. When the output is low, B2 supplies the current.

Notice that, when the duty cycle is 50% and the net power applied to the load is zero, both supplies (B1 and B2) are providing current, and the drained power is not zero. This fact should be considered when using this block.

Another point to be considered when using this circuit is that, when the applied power is zero (under the explained condition), the motor tends to vibrate at the oscillator frequency. This can be avoided by using a capacitor wired in parallel with the motor or by choosing a convenient frequency for the oscillator. This means that, depending on the motor, experiments must be conducted to find the correct value of C1.

Block 75   Power Anti-phase PWM Control Using the 555 IC

We can control dc motors with currents up to several amperes with the block shown in Figure 5.20. The NPN transistor conducts the current provided by B2 when the output of the 555 is at the high logic level, and the PNP transistor conducts the current provided by B1 when the output of the IC is at the low logic level.

The transistors are chosen as in previous projects (see the tables Chapter 4, Section 4.5). The reader can also use Darlington transistors in this control by increasing the value of the resistors at the transistors' bases from 1 kΩ to 4.7 or 10 kΩ.

Care must be taken when choosing the correct values for C1. This capacitor value can be established experimentally to match the characteristics of the motor, thus avoiding vibration when it is at the zero-power drive condition. A capacitor with values between 0.01 and 1 μF can be wired in parallel with the motor to avoid this vibration.

Remember that the great disadvantage of this kind of control is that, at the zero-power drive condition, the currents delivered by the power supplies are not zero, and a great deal of heat can generated in the motor. This kind of circuit is not suitable for the control of resistive loads such as heaters, SMAs, and others.

Block 76   PWM Control Using the LM350

The unusual configuration shown in Figure 5.21 uses an LM350 IC (linear voltage regulator) as a PWM control for dc loads up to 3 A. Small dc motors, lamps, heaters, and SMAs can be controlled by this circuit.

A variable duty-cycle oscillator using a 555 IC applies the squarewave signal to the adjustment input of the LM350. When the output of the oscillator is low, the

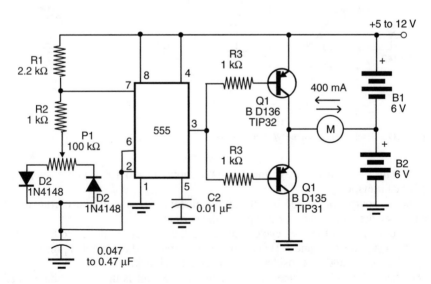

Figure 5.20   Block 75: anti-phase power.

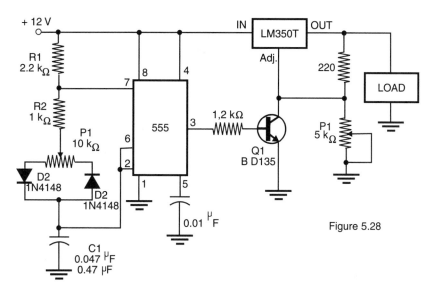

Figure 5.28

Figure 5.21   Block 76: PWM using the LM350.

voltage at the output of the LM350 falls to 1.25 V (the internal zener reference). When the voltage at the output of the 555 is high, the voltage at the output of the LM350 is given by the adjustment of P1. Controlling the duty cycle of the signal generated by the 555 IC, it is possible to control the power applied to the load.

The LM350 must be mounted on a heat sink. The frequency of the 555 must be determined experimentally to suit the motor. This circuit doesn't operate at high frequencies. The limit is given by the characteristics of the LM350T.

5.4   Additional Information

The following table provides information about some three-terminal, positive voltage regulators suitable for the applications described in this chapter. Figure 5.22 shows the pinout for these integrated circuits.

Type	Output Voltage	Current	Input Voltage	Remarks
LM7805	5 V	1 A	7 to 35 V	Fixed
LM7806	6 V	1 A	8 to 35 V	Fixed
LM7808	8 V	1 A	10 to 35 V	Fixed
LM7810	10 V	1 A	12 to 35 V	Fixed
LM7812	12 V	1 A	14 to 35 V	Fixed
LM7815	15 V	1 A	17 to 35 V	Fixed
LM7818	18 V	1 A	20 to 35 V	Fixed
LM7824	24 V	1 A	26 to 40 V	Fixed
LM217/317	1.2 to 37 V	1.5 A	40 V*	Variable
LM150/250/350	1.2 to 33 V	3 A	20 V*	Variable
LM138/238/338	1.2 to 32 V	5 A	40 V*	Variable

*Differential between input and output.

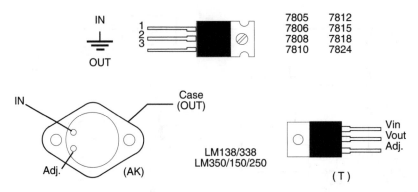

Figure 5.22   Pinout for various regulator configurations.

5.5   Suggested Projects

PWM or linear controls and H-bridges can be ganged together to form a complete control for dc motors in robotics or mechatronics applications. Some suggestions on how to accomplish this are the following:

- Blocks 58 and 95 can be adapted to form a unique Darlington PWM control.
- The blocks with relays can be used to cut the power of any H-bridge and/or PWM control when a low-current stand-by state is needed. Try to design a PWM control with an H-bridge with a zero current state using the blocks with relays.
- The above concept is valid if you want to add inertia or a delayed action. Combine the blocks with relays with PWM and H-bridges to add this new feature.
- An elevator with speed and direction control and inertia can be designed using all the blocks we have seen in the previous chapters.

5.6   Review Questions

1. How much power is converted into heat by a linear control when the input voltage is 12 V and the voltage in a 5 W load is 6 V?
2. What is a rheostat?
3. In a linear power control, a transistor operates like what kind of passive component?
4. Can a variable voltage regulator be used as a linear power control?
5. What is a constant current source?
6. What is duty-cycle?
7. In the anti-phase PWM, what is the power applied to a load when the duty cycle of the control signal is 50%?
8. Why is it not possible to apply 100% of the power to a load using a PWM control?

6

Power Control Using Thyristors

6.1 Purpose

Semiconductor devices of the thyristor family can be used for power control purposes. These include silicon controlled rectifiers (SCRs), triacs, silicon bilateral
switches (SBSs), silicon unilateral switches (SUSs), diacs, and others. This chapter
will show the reader how SCRs, triacs, and other thryristors can be used in robotics,
mechatronics, and other projects involving the control of high-power loads such as
motors, lamps, solenoids, and so on.

6.2 Theory

Thyristors are four-layer semiconductor devices intended for power control applications. The family of thyristors is composed of devices that can be used in both ac
and dc circuits.

The basic applications include both on-off control and the phase control where
the amount of power applied to a load can be adjusted in a linear range of values. In
the following paragraphs, we will see how some thyristors work and how they can
be used in robotics and mechatronics.

6.2.1 Silicon Controlled Rectifiers

The silicon controlled rectifier (SCR) is a four-layer device of the thyristor family.
The the symbol, structure, and equivalent circuit are shown in Figure 6.1.

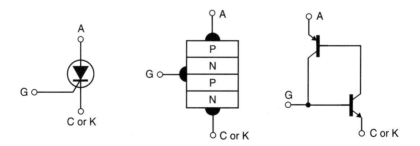

Figure 6.1   SCR: symbol, structure, and equivalent circuit.

The SCR, as the symbol suggests, can be compared to a diode that conducts the current in only one direction once triggered. To trigger on an SCR, it is necessary to apply a positive voltage to the gate. This voltage biases the NPN transistor of the equivalent circuit to the saturated state. This causes a feedback that forces the PNP transistor into the saturated state as well. Even when the trigger voltage disappears, the transistors remain on, due to feedback. To turn off the SCR, there are two possibilities:

1. Create a short with a switch between the anode and the cathode, making the voltage between these elements fall to zero. Pressing the switch momentarily turns the SCRs off.
2. Turn off the power to the circuit. When the power is turned on again, the SCR remains off if no trigger voltage exists in the gate.

Figure 6.2 shows the basic application of an SCR in a control circuit. S1 and S2 are used to turn off the circuit.

Common SCRs are very sensitive and can be triggered by currents as low as a few hundred microamperes. Examples include the TIC106, MCR106, IR106, etc. The trigger voltage for those devices is in the range of 1 to 2 V. The main current, between anode and cathode, is typically in the range of 100 mA to more than 100 A.

Using the SCR.   SCRs can be used in both ac and dc circuits. In dc circuits, we only must remember that the device remains on after the trigger signal is removed. In the ac circuits, the SCR turns off when the voltage crosses the zero point at the end of each semicycle as shown by Figure 6.3.

After an SCR is turned on, we must keep a minimal amount of current passing across it to keep it on. This current is called *holding current* and is in the range of a few milliamperes for common devices.

Another important point to be considered when using SCRs in low-voltage circuits is that a voltage drop of about 2 V is produced when it is on, as shown in Figure 6.4.

In some robotics and mechatronics applications, this voltage drop must be compensated by an increase in the power supply voltage. In high-voltage applications (plugged into the ac power line, for instance), this voltage drop can be ignored. This voltage also determines the amount of power converted into heat by the device when in operation.

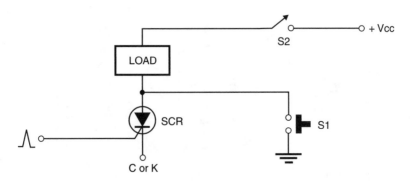

Figure 6.2   Circuit using an SCR.

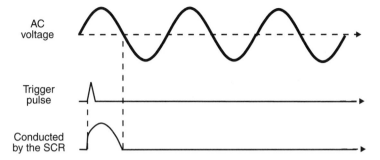

Figure 6.3   The SCR conducts on half-cycle when triggered.

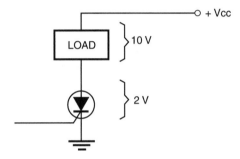

Figure 6.4   Voltage drop across an SCR.

6.2.2 Triacs

The triac is another semiconductor of the thyristor family, and it can be used to switch either ac or dc loads. The difference between the SCR and the triac is that the triac can conduct current in both directions. We can think of the triac as two SRCs wired in an anti-parallel mode with a common electrode for the gate. Figure 6.5 shows the structure and symbol of a triac.

The triac is found primarily in ac circuits, since it is a bilateral device. In dc applications, the SCR is preferred. Common triacs typically have trigger voltages in the range between 1 and 2 V and currents in the range between 10 and 50 mA. The main current can rise from 1 or 2 A to more than 100 A.

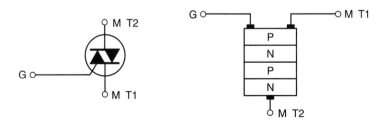

Figure 6.5   Triac, symbol and structure.

Using the triac.   Figure 6.6 shows the basic application circuit of a triac. The load is connected in series with main terminal 2 (MT2), and main terminal 1 is run to ground. There are four ways to trigger the triac, as shown in Figure 6.7. In basic applications, the trigger mode in the first quadrant is the preferred one, as it provides more sensitivity.

When turned on, a voltage drop of about 2 V occurs in the device. This voltage drop determines the amount of heat produced by the device (voltage drop × current = dissipation power).

6.2.3   Other Devices of the Thyristor Family

Many other devices of the thyristor family can be used in projects involving robotics and mechatronics. The following are the most important.

UJT.   The unijunction transistor (UJT) is a device that presents a negative input resistance. This characteristic makes it useful as an oscillator in low-frequency applications. Figure 6.8 shows the symbol, structure, and characteristic output of this device.

The UJT can be used as a relaxation oscillator in the configuration illustrated by Figure 6.9. This circuit can be used to produce sawtooth waves or pulses in the fre-

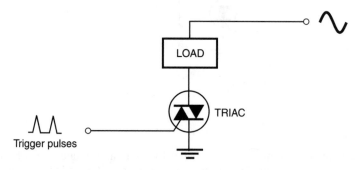

Figure 6.6   Using the triac.

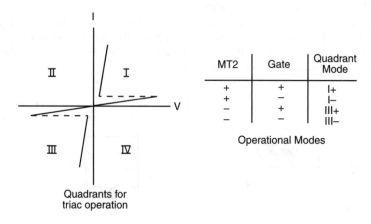

MT2	Gate	Quadrant Mode
+	+	I+
+	−	I−
−	+	III+
−	−	III−

Operational Modes

Quadrants for triac operation

Figure 6.7   Triac triggering modes.

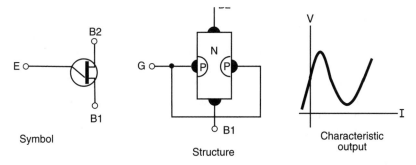

Figure 6.8   The UJT.

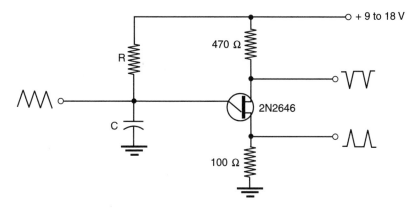

Figure 6.9   The relaxation oscillator.

quency range between a fraction of a hertz and 100 kHz. The most popular of the unijunction transistors is the 2N2646.

SUS.   SUS is the abbreviation for *silicon unilateral switch*. This device also shows a negative resistance characteristic that makes it useful as a trigger device for SCRs. Figure 6.10 shows the symbol of this device and how it can be used in circuits with SCRs.

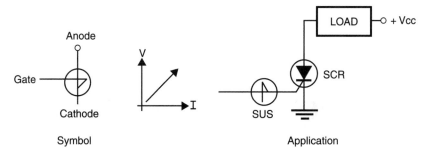

Figure 6.10   The SUS.

Diac.   The diac is a trigger device for triacs. This semiconductor device passes from the off state to the on state in a very short time when the voltage across it rises to a certain value (about 27 V for common types). As the diac triggers on with negative or positive voltages, it can be used to trigger triacs as shown in Figure 6.11.

In some cases, the diac and the triac are mounted inside the same package. The device formed by these two elements is also called a *quadrac*.

SBS.   The *silicon bilateral switch* is another trigger device for circuits using a thyristor. The symbol and appearance of this component are shown in Figure 6.12. The difference from this device and the diac is that the trigger voltage can be programmed by the control electrode.

Neon Lamp.   The neon lamp is not really a solid state device, of course. But this device, because of its negative resistance characteristic, is useful in many circuits as a trigger element. Figure 6.13 shows the symbol and appearance of a neon lamp.

The neon lamp passes from the off state to the on state, conducting current when the voltage across its terminals rises to about 80 V. In many applications, the neon lamp can replace diacs and SBS triggering thyristors such as triacs and SCRs.

Figure 6.11   The diac.

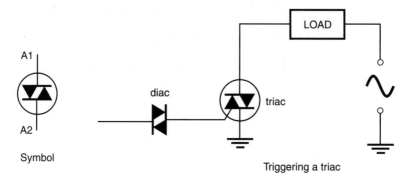

Figure 6.12   The SBS.

Figure 6.13   Neon lamp.

6.3　Basic Blocks Using SCRs

Now let's suggest to the reader some blocks with thyristors that can be used in projects involving robotics, mechatronics, and general automation. The blocks use common SCRs and triacs such as the ones in the TIC series. Variations in the component composition can be implemented to adapt these circuits to use other thyristors. The reader should feel free to make such experiments.

Block 77　Turning a Load On and Off with an SCR

The basic block shown in Figure 6.14 can be used to trigger loads on and off with currents up to 3 A. The switches can be replaced with sensors (magnetic or pendulum) in robotics and mechatronics applications. When S1 is momentarily closed, the SCR turns on and remains on even when S1 is released. To turn off the circuit, it is necessary to momentarily close S2.

Observe that the turn-on current is very low (determined by R1), but the turn-off current is high (the nominal current of the load). This fact must be considered when using when choosing the appropriate switches.

The reader must consider the voltage drop of about 2 V in the SCR when it is on. The value of R1 and R2 depends on the SCR. The values given in the block are recommended for any SCR in the 106 series. The circuit can be powered with voltages in the range between 6 and 150 V (see the suffix of the SCR).

The following table gives the recommended values for R1 and R2 according to the power supply voltage.

Power Supply Voltage	R1	R2
6 to 12 V	1 to 10 kΩ	150 Ω to 2.2 kΩ
12 to 24 V	4.7 to 47 kΩ	1 to 10 kΩ
24 to 48 V	10 to 47 kΩ	1 to 22 kΩ
48 to 100 V	22 to 100 kΩ	4.7 to 22 kΩ

An interesting application for this circuit is as a one-bit memory. An incandescent lamp is used as load. If the sensor is ever closed, the lamp remains on, indicating the condition.

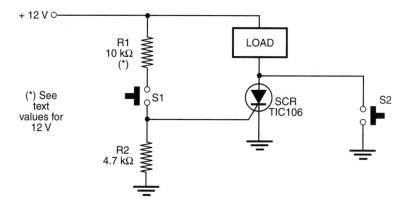

Figure 6.14　Block 77: turning a load on and off with an SCR.

Block 78   Delayed Turn-On Switch with SCR

High-sensitivity SCRs (TIC106, MCR106, and others) can be triggered from very low currents. This means that high-value resistances in RC circuits can be used to delay the turn-on point of an SCR.

Using a 100 kΩ resistor in the circuit shown by Figure 6.15, we can delay the turn-on instant of an SCR for up to several minutes. The resistor can be replaced with a 100 kΩ variable resistor in series with a 1 kΩ resistor if the reader needs to adjust the time delay. To turn off the circuit, S2 is used. The circuit can also be turned off by switching off the power supply voltage.

Block 79   Touch Switch Using an SCR

Even the low current passing across the fingers of a human is enough to trigger sensitive SCRs such as the TIC106 (see Chapter 10 for more details). The circuit illustrated by Figure 6.16 can be used to trigger loads up to 3 A by a single touch on two metal plates.

To turn off the circuit, S2 is pressed momentarily. The current in the sensor is less than 1 mA and is fixed by R1. The sensor is formed by two small, metal plates (1 ×

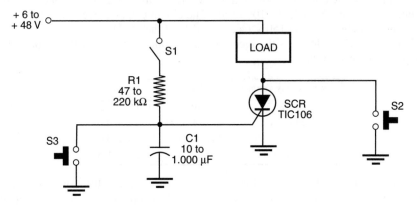

Figure 6.15   Block 78: delayed turn-on switch.

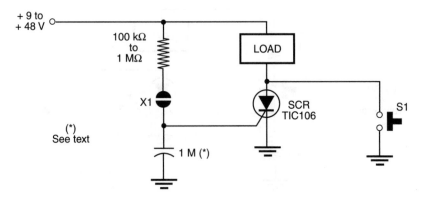

Figure 6.16   Block 79: touch switch.

2 cm), and the wire used to connect it to the circuit must be short to avoid hum. A 1 MΩ potentiometer between the gate and the ground can be added as a sensitivity control.

This configuration can be used with other resistive sensors, but the specific configurations are given in Chapter 10, where sensor blocks are described.

Warning: Don't power this circuit from the ac power line using transformerless power supplies. Touching the sensor can be a severe shock hazard.

Block 80   Triggering an SCR with Positive Pulses

Low-sensitivity SCRs can be triggered with positive pulses from TTL or CMOS logic using the block shown in Figure 6.17. R1 and R2 are chosen according to the current necessary to trigger the SCR.

The values shown in the circuit are for TIC126 SCRs. Although the current necessary to trigger this circuit is very low, the voltage isn't. The value of resistor R1 can also be increased to 1 MΩ, and R2 and R3 to 10 kΩ and 2.2 kΩ, if the SCR is the TIC106. In this case, the circuit will need less than 1 µA to trigger on.

Block 81   Triggering SCRs with Negative Pulses

The circuit shown in Figure 6.18 is compatible with TTL and CMOS logic, as it needs only a few microamperes to trigger the SCR on. The input must be kept at the high logic level to maintain the SCR in the off state. When the input passes to the low logic level for an instant, the transistor is cut off, and the SCR triggers by the current flowing across resistor R3.

The circuit is very sensitive, and SCRs as the TIC126 and other can be used. Diodes can be added to transform this circuit into AND and OR gates.

To turn the circuit off, the input must be returned to the high logic level, and the circuit power can be cut pressing a momentary contact switch wired between the anode and the cathode of the SCR.

Block 82   Crowbar Protection

This circuit is used to speed up the burnout of a fuse when the current rises above a predetermined value. The snappy action of the SCR speeds up the burnout process,

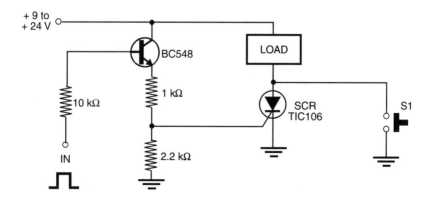

Figure 6.17   Block 80: triggering an SCR with positive pulses.

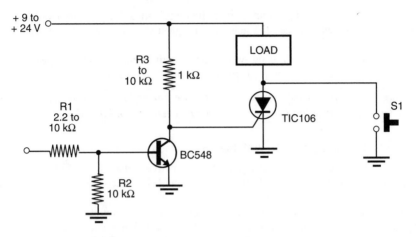

Figure 6.18   Block 81: triggering an SCR with negative pulses.

cutting off the current in a very short time interval. The circuit shown in Figure 6.19 is important for protecting sensitive circuits from shorts or large voltage changes.

The voltage drop across resistor R1 depends on the current. If the voltage drop rises to the trigger point of the SCR, the SCR turns on, putting the power supply in a short. This causes the fuse to blow, and the current flow is stopped.

The value of R1 depends on the current the designer specifies to blow the fuse. The value is found by the following formula:

$$R = \frac{V}{I} \qquad (6.1)$$

where   R = the resistance in ohms
V = SCR trigger voltage (typically between 0.8 and 1.2 V for the TIC106)
I = trigger current

Remember that the SCR must be able to tolerate the current necessary to blow the fuse.

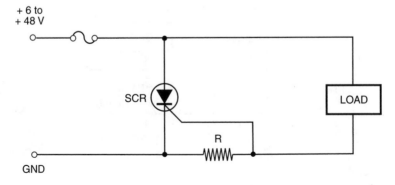

Figure 6.19   Block 82: Crowbar.

Block 83   Overcurrent Protection

This protection consists of a nondestructive circuit; no fuse is blown if an overcurrent current condition occurs. As the reader can see by Figure 6.20, a relay turns off the voltage applied to the load if the current rises above a predetermined value.

The relay stays on even when the overcurrent condition disappears. To reset the circuit, it is necessary to turn the power supply off and on or to connect a switch between the anode and the cathode of the SCR.

Remember that a voltage drop of about 2 V occurs across the SCR. We can compensate for this if the circuit operates with low voltages. Resistor R is calculated using the formula given in Block 82.

Block 84   R-S Flip-Flop Using an SCR

A common problem found in the previous blocks is that the SCR can be turned off only by a switch in parallel or by turning the power supply off and on. The switch, if installed, must be able to handle all the SCR current, which can be inconvenient in many cases.

The ability to turn off the SCR via a signal or using low-power switches can be important in many robotics and mechatronic projects. The block shown in Figure 6.21 can accomplish that.

The SCR is turned on by momentarily pressing S1 (which can be replaced with any sensor). To turn off the circuit, it is sufficient to press S2 momentarily.

When SCR1 is on, the capacitor remains charged, since the side of SCR1 is at ground potential, and the side of SCR2 is at the positive voltage.

When S2 is pressed, SCR2 turns on, and the capacitor is discharged. The discharge current acts as a short between the anode and cathode of SCR1, turning it off. To turn it on, press S1. Now, SCR1 turns on, and the discharge current of C1 acts as a short for SCR2, turning it off.

The value of the capacitor depends on the application and, in general, is in the range between 1 and 10 µF. Polarized types need not be used for this function. The same principle used to turn off this circuit can be used in other blocks.

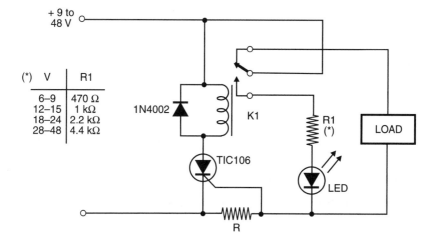

Figure 6.20   Block 83: overcurrent protection.

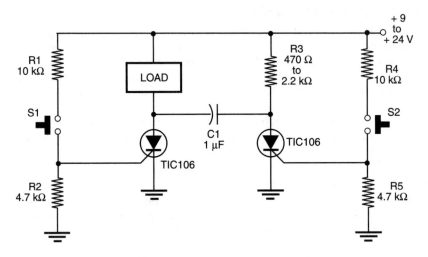

Figure 6.21   Block 84: R-S flip-flop using an SCR.

6.3.1   SCRs in AC Circuits

The SCRs can be used in ac circuits, too. We just have to remind the designer that, when powered from an ac source, the SCR turns off when the voltage crosses zero at the end of each semicycle.

In the applications described in the following blocks, the circuit can be powered from the secondary of a transformer and, in some cases, directly from the ac power line. It is important to observe only that:

1. The SCR must be able to handle the peak voltage and current present in the circuit.
2. In applications powered from the ac power line, care must be taken to avoid creating shock hazards.
3. If only one SCR is used, only half of the semicycles are conducted.

Block 85   Simple AC Switch

The circuit shown in Figure 6.22 can be used to turn an ac load on and off. Notice that the circuit conducts only the positive semicycles of the power supply voltage. This means that approximately half of the total power is applied to the load. For low-power loads, it is possible to compensate for this loss of power in the load by connecting a large value capacitor in parallel.

Block 86   Full-Wave AC Switch (I)

This block eliminates the inconvenience of Block 85 (loss of half of the power), as shown in Fig. 6.23. The full bridge in the input enables the operation of the SCR with both the negative and positive semicycles of the power supply voltage. The diodes must be rated for the voltage and current required by the load. The switch can be replaced by a sensor.

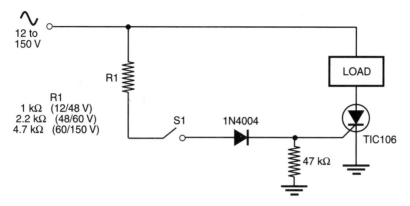

Figure 6.22   Block 85: simple ac switch.

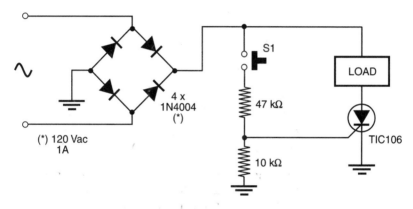

Figure 6.23   Block 86: full-wave ac switch I.

Block 87   Full-Wave AC Switch (II)

The block shown in Figure 6.24 shows another way to connect the load in a full-wave switch using an SCR. The circuit operates in the same manner as the previous one.

Block 88   Dimmer and Speed Control

The block shown in Figure 6.25 describes a more traditional configuration for an SCR as a phase control for ac loads. This circuit can be used to control the brightness of an incandescent lamp, the temperature of a heater, or the speed of an ac motor.

When a half cycle of the ac power voltage begins, the voltage rises at the same time that C1 charges across potentiometer P1 and resistor R1. The time constant of these components is chosen to fall within the time interval corresponding to a half cycle. This means that the SCR will trigger at a point of the half cycle that depends of the adjustment of P1. If P1 is adjusted for low resistance, the time constant is short, and the voltage to trigger the SCR is reached quickly. The SCR turns on at the beginning of the half cycle, allowing almost all of the power to go to the load.

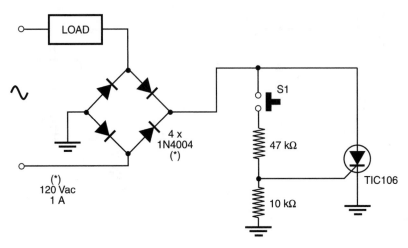

Figure 6.24   Block 87: full-wave ac switch II.

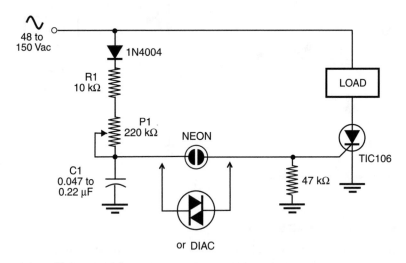

Figure 6.25   Block 88: ac dimmer and speed control.

On other hand, if P1 is adjusted for a high resistance, the voltage to trigger the SCR on will be reached only at the end of the half cycle, and the power that passes to the load will be low. Notice that the SCR turns off again at the end of each half cycle of the power supply voltage.

From the lowest resistance point to the highest, it is possible to control the power applied to the load over a large power range. Although this kind of circuit is very efficient when controlling resistive loads, it also is inconvenient in some ways:

1. The fast on and off switching of the SCR can generate noise. The noise can travel through the power supply line or be radiated in the form of electromagnetic waves. In many cases, you will need to use filters to avoid problems.
2. Only the positive half cycles are controlled.

If the circuit is powered from the ac power line, some kind of device can be added to speed up the triggering process of the SCR. This device can be a neon lamp or a diac in the gate, as shown by the figure.

Block 89   Full-Wave Dimmer

The inconveniences caused by the half-wave operation of the previous block can be eliminated using the circuit shown in Figure 6.26. The operating principle of this circuit is the same as that of the previous block.

Block 90   Dimmer Using UJT and SCR

The circuit shown in Figure 6.27 can be used as a linear power control for loads in the range between 12 and 48 V. Depending on the power supply voltage, capacitor

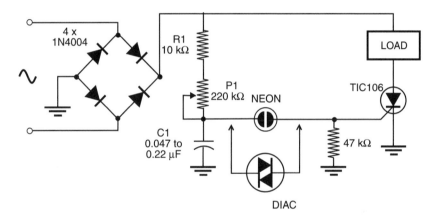

Figure 6.26   Block 89: full-wave dimmer.

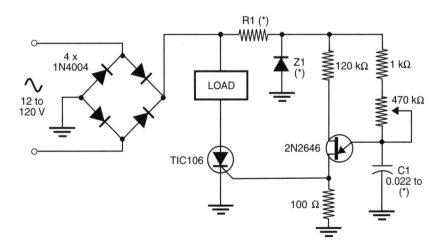

Figure 6.27   Block 90: dimmer using UJT and SCR.

C1, resistor R1, and the zener must be replaced with components with different values. The following table gives recommended values for these components:

Power Supply voltage (V)	Z1	C1	R1
12 to 15 V	9 V × 400 mW	0.022 µF	470 Ω
15 to 24 V	12 V × 1 W	0.033 µF	680 Ω
24 to 30 V	15 V × 1 W	0.047 µF	1 kΩ
30 to 48 V	22 V × 1 W	0.068 µF	1.5 kΩ

Because of tolerance variations among the components, particularly among capacitors, you may need to experiment with each circuit to find the best values. The values suggested in the table are average.

6.3.2   Blocks Using Triacs

The triac is recommended for special for applications where high-power loads are connected directly to the ac power line. Since, in our case, the recommendation is to power these circuits from low-voltage sources, we do not expect the designer to employ many blocks that use triacs. Therefore, only a few basic triac-based blocks are included here. Starting with these blocks and using them as analogies, the designer can extend many applications to ac loads if desired.

Block 91   High-Power AC Switch

High-power ac loads can be controlled by a switch (or sensor) using the block shown in Figure 6.28. For the 127 Vac power line, R1 is a 470 Ω × 1 W resistor. For the 220 or 240 Vac power line, use a 1 kΩ × 1 W resistor. The current in the switch depends on the triac and rises to about 20 mA in such types as the TIC226. Note the suffix of the triac, which refers to the ac power line voltage.

Block 92   High-Power Dimmer

The circuit shown in Figure 6.29 is a linear power control that can be used with both resistive loads (lamps or heaters) and inductive loads (motors and solenoids). The capacitor value depends on the ac power line, and the triac is chosen according to the load current. Any diac can be used in this project. In some cases, the diac can be replaced with a neon lamp.

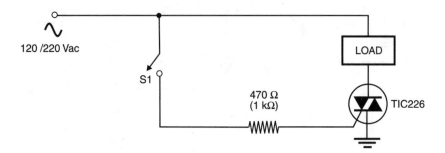

Figure 6.28   Block 91: high-power ac switch.

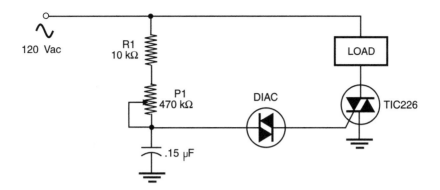

Figure 6.29   Block 92: high-power dimmer.

6.4   Additional Information

6.4.1   Care when Using Inductive Loads

When using inductive loads, it is necessary to add a diode in parallel as shown in Figure 6.30. If this diode is not used, when the load is turned on, the voltage pulse produced in the coil of an inductive load can turn the SCR off.

The on and off switching process that occurs when small dc motors function can make it necessary to use some additional parts to avoid turning off of the SCR after it is triggered by short pulses. Figure 6.31 shows some solutions to this problem.

In Figure 6.31a, a resistor intended to maintain the holding current in the circuit avoids the SCR turnoff when the current falls to zero as the coils are switched. In 6.31b, we use a diode, and in 6.31c a capacitor. The reader must conduct experiments to find the correct solution for each case.

6.4.2   Characteristics of Some Common SCRs and Triacs

SCRs

Type	Voltage (Vdrm)	Current (dc)	Holding Current (max)	Trigger Current (typ)	Trigger Voltage (typ)
TIC106-Y	30 V	3.2 A	5 mA	60 µA	0.6 V
TIC106-F	60 V	3.2 A	5 mA	60 µA	0.6 V
TIC106-A	100 V	3.2 A	5 mA	60 µA	0.6 V

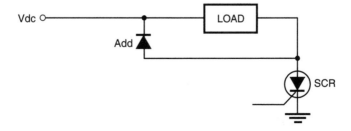

Figure 6.30   Adding a diode in parallel with inductive loads.

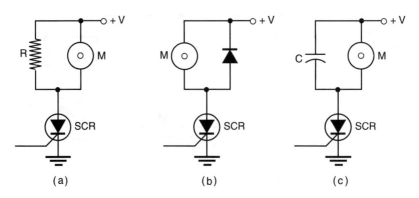

Figure 6.31   Controlling dc motors.

SCRs (continued)

TIC106-B	200 V	3.2 A	5 mA	60 μA	0.6 V
TIC106-D	400 V	3.2 A	5 mA	60 μA	0.6 V
MCR106-1	30 V	4.0 A	5 mA	200 μA	1.0 V
MCR106-2	60 V	4.0 A	5 mA	200 μA	1.0 V
MCR106-3	100 V	4.0 A	5 mA	200 μA	1.0 V
MCR106-4	200 V	4.0 A	5 mA	200 μA	1.0 V
TIC126-B	200 V	8 A	70 mA	5 mA	0.8 V
TIC126-D	400 V	8 A	70 mA	5 mA	0.8 V

Triacs

Type	Voltage (Vdrm)	Current	Holding Current (max)	Trigger Current (max)	Trigger Voltage (max)
TIC226-B	200 V	8 A	60 mA	50 mA	2.5 V
TIC226-D	400 V	8 A	60 mA	50 mA	2.5 V

6.5   Suggested Projects

The blocks included in this chapter can be used alone or ganged with other blocks to create many projects in mechatronics or robotics. Some suggestions are as follows:

- Block 77 or 78 can be used in security systems for automatic devices that use a relay as a load. The relay will trigger some sort of warning system when a dangerous condition is detected by the sensor (S1). The blocks can also be used as an emergency switch.
- Blocks 82 and 83 can be used to detect when a motor is stalled in an overload condition. In the first case, the fuse blows, and in the second, the motor is turned off.

6.6   Review Questions

1. Is the SCR a full- or half-wave device?

2. What is needed to trigger an SCR off when it controls a dc load?
3. What is a quadrac?
4. What is the holding current of an SCR?
5. How can an SCR can cause interference in radio receivers while controlling ac loads?
6. What is name of the configuration used to speed up fuse burnout in a protective circuit?
7. What does the suffix "D" mean in the SCR TIC106-D?

7

Solenoids, Servomotors, and Shape Memory Alloys

7.1 Purpose

The purpose of this chapter is to introduce the reader to the use of shape memory alloys (SMAs), solenoids, and servomotors (often just called *servos*) in mechatronics and robotics projects. The basic idea is to show how these devices can be used as alternative solutions for the generation of force and movement. The reader will also see how many blocks shown in the previous chapters can be used to drive these devices and how to design their circuits for practical purposes.

7.2 Theory

Mechanical power sources in practical robotics and mechatronics projects are not limited to dc and ac motors. In many cases, the desired movement or mechanical force can be advantageously provided by other categories of devices. In this chapter, we will study three of these devices in particular: SMAs, solenoids, and servomotors. The use of any of these devices depends on many factors that must be considered for each project. Knowing how they work and what they can do, the reader will be able to make the correct choice and design the elements of the drive circuit.

7.2.1 Shape Memory Alloys

Shape memory alloys are materials that have the ability to return to a predetermined shape when heated. When an SMA is cold (i.e., below a temperature called the *transformation temperature*), it displays a low yield strength and can be deformed quite easily into a new shape that is retained by the material. However, if the material is heated above the transformation temperature, it undergoes a change in atomic structure that causes it to return to its original shape. If the SMA encounters any resistance during this transformation, it can generate large forces. In robotics and mechatronics, an SMA wire can be used as an *electronic muscle,* as shown in Figure 7.1.

When heated by a current, the SMA length changes, and the force produced in this process can be used to move an arm, leg, lever, or other moving part of a mechanism. The most common SMA material is a nickel-titanium alloy called *nitinol* or *flexinol*. This material has good electrical and mechanical properties, long fatigue life, and high corrosion resistance.

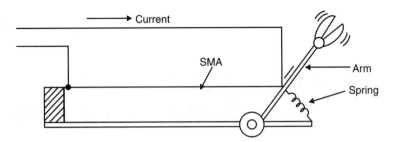

Figure 7.1   Using an *electronic muscle.*

A nitinol wire 0.02 inches in diameter can lift as much as 16 lb (8 kg). A flexinol wire only 100 microns in diameter can be deformed by 28 g of weight and produce 150 g of recovery force when exposed to a 180 mA current.

The typical SMA at room temperature is easily deformed by a small force. However, when conducting an electric current, it changes to a much harder form that returns to the original shape. A stretched SMA wire, as it shortens, produces a usable amount of force. An SMA wire can be stretched by as much as 8% of its original length.

Of course, SMAs are not recommended for all the applications. One must take into account the forces, displacements, and other conditions required for a particular application. The most important use of SMAs is in small actuators, where mechanical devices such motors and solenoids are not easy to implement.

One of the dealers of nitinol SMA is TiNi Alloy Company (http://www.tini.com), which sells some types of Nitinol wire and can provide information about its use. [See Section 7.4, Additional Information, for details on how to obtain SMAs for experiments. Another reputable dealer of nitinol SMAs is the Robot Store (http://www.RobotStore.com).]

Properties of Nitinol SMA

- Melting temperature: 1240 –1310°C
- Resistivity (high-temperature state): 82 Ω-cm
- Resistivity (low-temperature state): 76 Ω-cm
- Thermal conductivity: 0.1 W/cm-°C
- Ultimate tensile strength: 750–950 Mpa or 110–140 ksi
- Typical elongation to fracture: 15.5%
- Energy conversion efficiency: 5%
- Work output: 1 Joule/gram
- Available transformation temperatures: –100 to +100°C
- Approximate Poisson's ratio: 0.3

How To Use SMAs

The use of SMA wires in robotics and mechatronics is simple: the wire is used to move mechanisms (arms, grippers, levers, etc.,) or employed in electric pistons. Electric pistons, as shown in Figure 7.2, can be used as solenoids but act by the contraction and distension of SMAs rather than by the force of magnetic fields. The basic application is shown in Figure 7.3.

Figure 7.2   An electric piston.

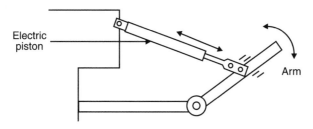

Figure 7.3   Basic application.

The length of the wire multiplied by the resistance per meter gives the total resistance of the muscle. Applying Ohm's law, we can determine the voltage that should be used to give the amount of current required by the specific type.

In the inverse manner, we can determine the length from the voltage and the current. For instance, if we want to power a muscle using a Flexinol 250 SMA from a 12 V power supply, what is the length to be used?

$$V = 12 \text{ V}$$
$$I = 1 \text{ A (from the specifications)}$$
$$R = 12/1 = 12 \ \Omega$$

The length therefore is

$$L = 12/20 = 0.6 \text{ m or 60 cm}$$

7.2.2   The Solenoid

If you want to create a small displacement of a moving mechanical part, consider the use of a solenoid. A solenoid is formed by a coil in which a moving magnetic core can slide as shown in Figure 7.4.

When a current passes through the coil, the magnetic field produces a force that attracts the moving core. If this moving core is coupled to any mechanism, the force that created by this process can be used to move it.

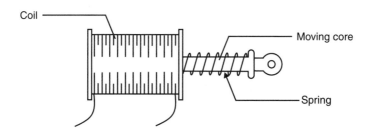

Figure 7.4   The solenoid.

Since the force that attracts the core is present only while the current passes through the coil, and no force is produced when the current is off, a spring or other mechanism must be added to return the core to its original position. Figure 7.5 shows some applications of solenoids in simple mechanisms.

In Figure 7.5a, we show a simple arm using a solenoid. In 7.5b, we show a steering system for a robot using a solenoid. In 7.5c, we illustrate a system that opens a door when a solenoid is switched on.

Using a solenoid.   Simple solenoids can be wound by the designers or bought ready to be used from some dealers. The amount of force that a solenoid can provide to a mechanism depends on two factors: the current across the coil and the number of turns in the coil.

The current is determined by the ohmic resistance and the voltage applied to the device. This means that the main specifications of a solenoid include the voltage and the current, and sometimes the electromotive force. In some cases, instead the current, the ohmic resistance is given. In this case, you just need to divide the voltage by the ohmic resistance to find the nominal current.

Solenoids reassemble relays in some ways: if desired, you can apply a voltage higher than the nominal voltage for few seconds to obtain short bursts of higher power. You can also apply lower voltages if you want less power.

The most common types of solenoids are designed to be driven from dc supplies in the range from 10 A down to 2 or 3 A, with voltages in the range from 1 to 48 V. But some solenoids are designed to be driven from the ac power line.

7.2.3   The Servomotor

Common dc and ac motors are *open loop* designs. Drawing current from a power supply, they perform their action by moving a shaft or a gearbox. Servomotors, in

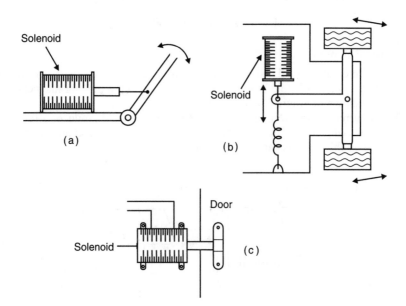

Figure 7.5   Using solenoids.

contrast, are *closed loop* devices. When provided with a control signal, the servomotor adjusts its action to match the signal. If the signal changes, the servomotor compensates. Servomotors are dc gear motors that include a closed loop position control. The shaft of the motor can be positioned or rotated through 180° as shown by Figure 7.6.

Servos are used in the hobby radio control (R/C) market for controlling model airplanes, cars, boats, and robots. The reader can find servos in many sizes according to the power required and the tasks for which they are intended. Commercial servos can be used with many of our blocks in robotic applications.

The standard servo has three wires: two for power (4 to 6 V) and ground, and the third for control. In common types, the signal is a variable-width pulse. The neutral position corresponds to a pulse of about 1.5 ms, sent at intervals of 20 ms (50 times per second). The control range (−90 to + 90°) corresponds to the pulse range of 1 to 2 ms.

If desired, you can build your own servo. Figure 7.7 shows a simple project. This servo is controlled by voltage instead of pulses (but we can adapt it to operate with pulses). The circuit schematic is shown in Figure 7.8.

In Figure 7.7, a gearbox is adapted to use a worm screw that moves a nut. The nut is coupled to a slide potentiometer that gives electric feedback to the circuit. When you apply a voltage to the input of the circuit (signal), the motor moves until the potentiometer reaches a position at which the feedback voltage is equal to the input voltage (zero output). If you change the voltage in the input, the motor will be powered until the potentiometer reaches a new position where the output is again zeroed.

Figure 7.6   A servomotor.

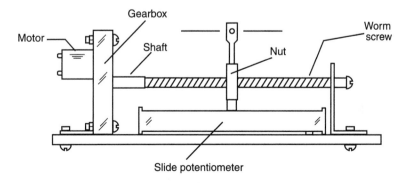

Figure 7.7   A home-made servo.

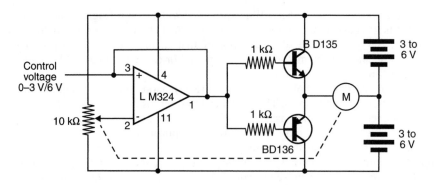

Figure 7.8   Circuit for the home-made servo.

7.3   Practical Blocks

Many basic blocks can be used in projects that use electronic muscles (based on SMAs), solenoids, and servos. Many of the ones referenced in this chapter were described in the previous chapters and have been recommended for the control of motors and other loads.

7.3.1   Blocks for Electronic Muscles (SMAs)

The electronic muscles (SMAs) and pistons can be thought of as resistances, since they have an ohmic resistance that changes with the temperature. This means that simple circuits used to drive heaters, lamps, and even dc motors can also be used. The following are some suggestions for practical applications.

Block 93   Simple Drive Circuit for SMAs

The block shown in Figure 7.9 illustrates the simplest way to drive an SMA. The voltage must be calculated according to the SMA (length, current, and resistance by meter) and the power supply must be able to supply the necessary current.

The SMA contracts when switch (S1) is closed and distends when the switch is open. The switch can be replaced with relays or sensors. Many of the previous blocks that use relays can be coupled to an SMA by adding a block like this one.

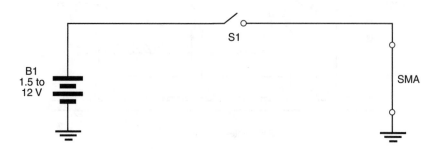

Figure 7.9   Block 93: simple SMA driver.

Block 94   Rheostat for SMA

The SMA requires a specific minimum amount of current to be heated to its transition temperature. You can determine this current using the circuit shown in Figure 7.10. This circuit can be connected to any power supply or batteries from 3 to 12 V and can produce currents up to 3 A with the recommended transistor. The transistor can be installed on a heat sink. For currents up to 2 A, the transistor can be replaced with a TIP32, and for currents up to 1 A with a BD135.

Block 95   Constant Current Source

The resistance of an SMA changes when it is heated, as we noted in Section 7.2.1, and this means that the current sourced by the power supply also changes. This variation can cause overheating or other problems. If the SMA is powered from a constant current source, secure operation can be achieved.

The block shown by Figure 7.11 can be used to maintain constant current across an SMA in a range from a few milliamperes to 3 A. The resistor, according to the desired current, is calculated by the following formula:

$$R = \frac{1.25}{I}$$

where   R = resistance of R in ohms
          I = desired current in the SMA

Other constant current sources described in this book can be used. In particular, we recommend Block 67, which uses a common bipolar transistor for the task.

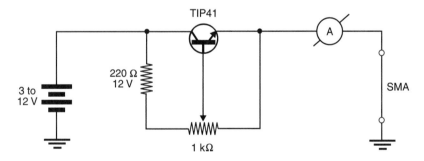

Figure 7.10   Block 94: rheostat for SMA.

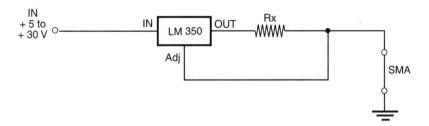

Figure 7.11   Block 95: constant current source.

Block 96   NPN Transistor Drive for SMA

Figure 7.12 shows a simple block using an NPN transistor driving an SMA (electronic muscle) up to 2 A. This circuit needs 20 mA to drive an 1 A SMA and less to drive lower-current muscles. TTL and CMOS outputs can drive low-power SMAs using this block. For higher gain, driving loads up to 1 A from TTL or CMOS outputs, you can replace the transistor with a BD135 or BD137. In any case, the transistor must be mounted on a heat sink.

Block 97   Using a Bipolar PNP transistor

The circuit shown in Figure 7.13 turns on the load (SMA muscle) when the input goes to the low logic level or to the ground. Any PNP power transistor can be used (see Chapter 3 and the circuits for more data about the transistor). This block, when using a BD136, needs only 2 mA to drive 500 mA SMA.

Block 98   Driving SMA from Power MOSFETs

Power MOSFETs are excellent devices for driving an SMA. A block using this kind of transistor is shown in Figure 7.14. See Chapter 3 to find a suitable power MOSFET for your application as appropriate to the input signal you intend to use. The gate resistor is used if a current limit is necessary. The transistor can be mounted on a heat sink.

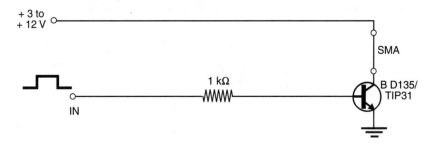

Figure 7.12   Block 96: NPN driver for SMA.

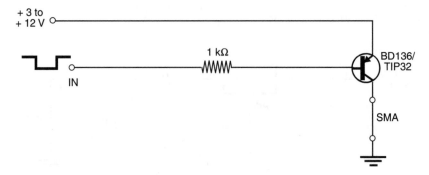

Figure 7.13   Block 97: Using a PNP transistor.

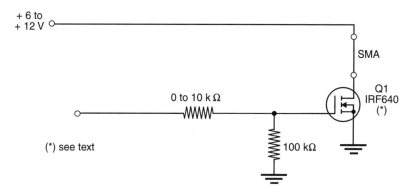

Figure 7.14   Block 98: driving SMAs from MOSFETs.

Other Blocks.   Any block described in this book for power switching can be used with an SMA. Remember that the SMA does not allow any kind of linear power control, as it operates only in a "yes/no" mode when it passes the transition temperature. PWM and linear blocks are not suitable for the control of electronic muscles.

In particular, we suggest using these blocks:

- Blocks 32 and 33: power control using NPN and PNP transistors
- Block 37: using a power MOSFET
- Blocks 39–41: delayed controls
- Blocks 77 and 78: blocks using SCRs
- Block 79: touch control
- Blocks 80 and 81: more blocks using SCRs
- Block 84: flip-flop using SCRs suitable for SMA applications
- Blocks 85–87: SCRs in dc circuits
- Blocks 88 through 90: more SCR blocks

7.3.2   Blocks for Solenoids

Many of the blocks used to drive SMAs and dc motors can be used with solenoids. The solenoid can considered as an inductive load requiring an amount of current that depends on the power needed to move a mechanical part. The following are some basic blocks, but we also will indicate previous blocks that can be used for the same task.

Block 99   Turning a Solenoid On and Off

The basic block recommended for turning a solenoid on and off is the one shown by Figure 7.15. There isn't much to say about it, as it is the same as recommended by Block 1. All motion control blocks shown in Chapter 2 can be used with solenoids.

Block 100   Determining the Force of a Solenoid

The circuit shown by Figure 7.16 can be used to determine the force of a solenoid and the current needed for it. The circuit is a simple rheostat. The designer can make

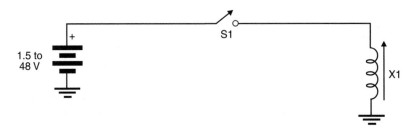

Figure 7.15   Block 99: controlling a solenoid.

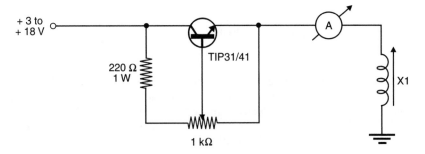

Figure 7.16   Block 100: determining the force of a solenoid.

a table noting all the values of current and voltage and the correspondent force measured by a dynamometer in the arrangement shown in Figure 7.17. Any other rheostat described in Chapter 3 can be used to control the force of solenoids.

Double Solenoid Actuators.   A simple arrangement of two solenoids is used in many direction controls and actuators as shown in Figure 7.18. When solenoid X1 is on, the robot turns to the left, and when the solenoid X2 is on, the robot turns to the right. Other applications include mechatronic actuators, such as the one shown in Figure 7.19, that can be used to move an object.

As the reader can see, in this application, only one solenoid can be on at a time. The following blocks are circuits to drive two solenoids in this type of application.

Block 101   Intelligent Control for Two Solenoids

The circuit shown in Figure 7.20 avoids the forbidden condition in which the two solenoids are activated at the same time. When the input is at the high logic level or re-

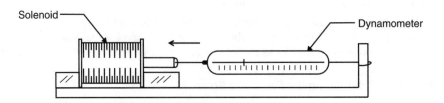

Figure 7.17   Measuring the force of a solenoid.

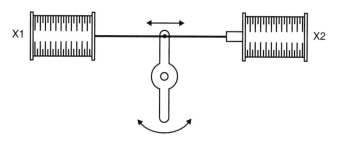

Figure 7.18   A double solenoid activator.

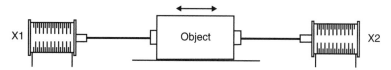

Figure 7.19   Moving an object with two solenoids.

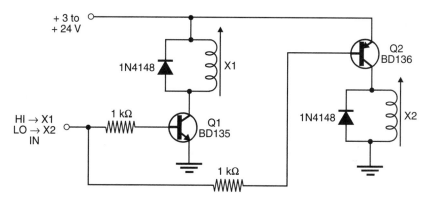

Figure 7.20   Block 101: intelligent control for two solenoids.

ceives a signal of up to 1 V, solenoid X1 is on and X2 is off. When the input is set to the low logic level or to the ground, solenoid X1 is off and X2 on.

The transistors are chosen according to the current required by the solenoids. Block 3 shows how to find the correct transistor for each application. The base resistor can also be altered in the range between 1 to 10 kΩ according to the gain of the transistors installed and the current required by the solenoids.

Block 102   Intelligent Solenoid Control Using Darlington Transistors

The same configuration described in the previous block, but using NPN Darlington transistors, is shown in Figure 7.21. This configuration needs less current to be driven. Only a few milliamperes are need to drive solenoids with currents up to several amperes, depending on the transistor (see Chapter 3 to choose the transistor). In particular, we recommend this configuration to drive high-current solenoids from TTL or CMOS outputs.

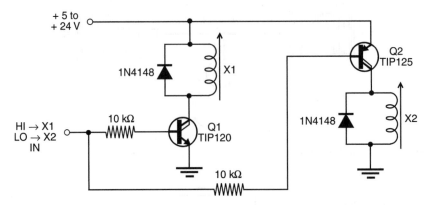

Figure 7.21   Block 102: intelligent block using Darlingtons.

Note: Equivalent configurations using a PNP bipolar transistor and a PNP Darlington transistor can be easily implemented, based on the circuits shown in Blocks 31 and 35. Configurations using two transistors can also be derived from Blocks 36 and 37.

Block 103   Intelligent Control Using CMOS Logic

In particular, the block shown in Figure 7.22 is recommended when the solenoids are controlled by CMOS logic. You can replace the IC with the equivalent TTL function. When the input is at the high logic level, solenoid X2 is on and X1 is off. At the low logic level, solenoid X2 is off and X1 is on.

When using the BD135 solenoids, up to 1 A can be driven. The reader must experiment with resistor values if other transistors are used. Darlington transistors can be used if high-power solenoids are installed. You can replace the resistors with 10 kΩ units in this case.

Block 104   Intelligent Circuit Using CMOS IC and Power CMOS

The circuit shown in Figure 7.23 is ideally suited to be driven from CMOS logic. Any CMOS function that can be used to invert the logic is suitable for this application. The power MOSFETs are chosen according to the current requirements of the solenoids. Operation is the same as in previous blocks.

Block 105   Current Sensing (I)

The amount of current passing across a solenoid depends on the position of the core. When operating with dc or pulses, the current changes according to the force, and the changes can be used to drive some external circuit. The circuits shown by Figure 7.24 can be used for current sensing in many applications.

In Figure 7.24a, we have direct sense. The output is 0.1 V per ampere. In 7.24b, we have the same circuit when the solenoid is driven by a transistor (bipolar, Darlington, or power CMOS). The value of Rx can be increased according to the application, but remember that, as the value becomes higher, so does the voltage drop in the component and the amount of heat it generates.

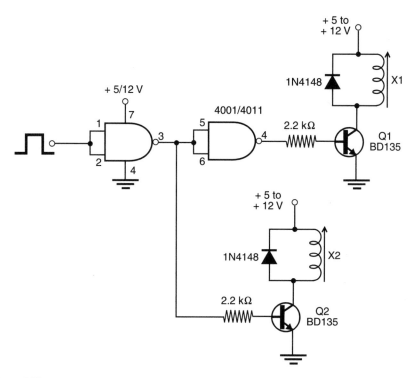

Figure 7.22　Block 103: using CMOS logic.

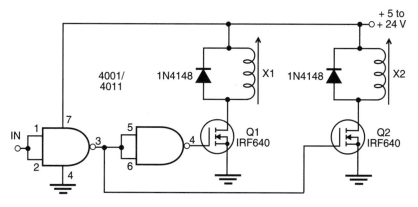

Figure 7.23　Block 104: CMOS intelligent control.

Block 106　Current Sensing (II)

The block shown by Figure 7.25a uses an operational amplifier to boost the voltage sensed by the circuit. In 7.25b, we show how to use an SCR to trigger some external load when the current in a solenoid rises above a predetermined value. The value of Rx is calculated as in Block 107.

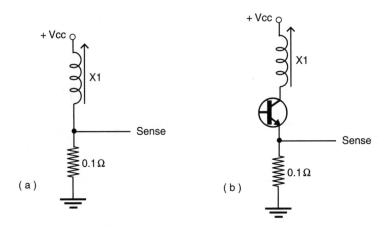

Figure 7.24 Block 105: current sensing I.

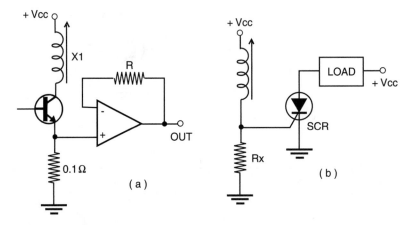

Figure 7.25 Block 106: current sensing II.

Other Blocks.   Many blocks described in previous chapters can be used with sole-noids in the following ways:

- Blocks 3 through 7: all these blocks are recommended for selective control of so-lenoids and solenoids in ac circuits.
- Blocks 8 through 13: these blocks can be used to control to power applied to a so-lenoid and also to provide inertia.
- Blocks 16 and 17: current sensing and control can be added with these blocks.
- Blocks 21 through 26: all of the blocks using relays can also control solenoids.
- Blocks 30 through 44: the blocks using transistors to control power loads can also be used to drive solenoids.
- Blocks 63 through 66: any of these blocks can be used to control the current across a solenoid.

- Blocks 67 through 69: constant current sources can be used with solenoids to keep the applied power constant.
- Blocks 70 through 73: PWM controls can be used with solenoids.
- Blocks 77 through 81: blocks using SCRs can control solenoids too.
- Blocks 84 through 92: ac and dc solenoids can be controlled by the circuits shown in these blocks. The flip-flop applications are important in many cases.

7.3.3　Blocks for Servos

Since servos can operate directly using signals that are generated by sensor and control circuits, few blocks can be suggested in this area. The following blocks are suggested for the control of servos used in R/C and other applications.

Block 107　Servo Control Using DC Motors

The circuit illustrated by Figure 7.26 is suggested by National Semiconductor and can control a home-made servo using a motor with currents up to 3 A. The basis is two LM675 power operational amplifiers wired as comparators. When you move P2 (the control), the motor will be energized to actuate on P1 until the voltage applied to the circuit reaches the same value of the voltage applied by P2. Observe that P1 must be ganged to the gearbox were the motor is installed. The circuit can operate with voltages starting at 6 V and can be adapted to use other operational amplifiers according to the motor requirements.

The same block can be used with external signals from sensors. It is sufficient to remove P2 and apply the signal directly to the 20 kΩ resistor. The voltage of this sig-

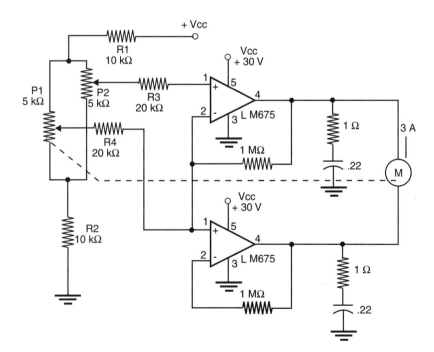

Figure 7.26　Block 107: servo control using a dc motor.

nal must be in the same range determined by the voltage divider formed by R1/P1 and R2.

Block 108   Control for R/C Standard Servos

The circuit shown in Figure 7.27 produces variable-width pulses for the control of commercial R/C servos. The variable resistors P1 and P2 must be adjusted to the range of width and separation needed by the servos. Once you have adjusted the circuit, you can alter the width of the pulse by using a sensor in place of the 330 Ω resistor, or even an external control circuit.

7.4   Additional Information

Readers who want to conduct experiments with SMAs can find some dealers who sell SMAs with different diameters. Samples with lengths starting at 1 m can be bought at low prices, and there are even kits for complete robots and mechatronics projects. One dealer of nitinol SMA is TiNi Alloy Company (http://www.tini.com), which sells several types of Nitinol wires and provides application data. We suggest that the reader also take a look at the Robot Store of Mondo-Tronics at (http://www.robotstore.com). This company has Flexinol kits and robots in wide variety of sizes, including all the information the designer needs to work with them. Table 7.1 shows the properties of the flexinol "muscle wire" from Mondo-Tronics.

7.5   Suggested Projects

The SMA (electronic muscle) can exist in only two conditions, since the recovery force cannot be controlled by the amount of current across it. This means that electronic muscles are suitable only for applications in which you don't need linear control of the force. The following are some suggested projects:

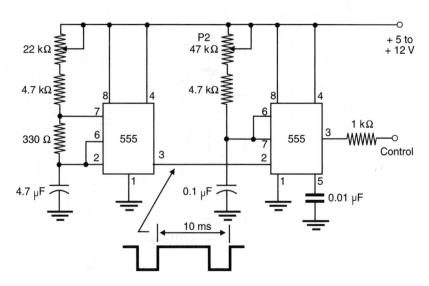

Figure 7.27   Block 108: control for standard R/C servos.

Table 7.1   Flexinol Properties

Wire Name	Diameter (microns)	Linear Resistance (Ω/m)	Typical Current (mA)	Deform. Weight* (grams)	Recovery Weight* (grams)	Typical Rate† (LT/HT)‡
Flexinol 037	37	860	30	4	20	52/69
Flexinol 050	50	510	50	8	35	46/67
Flexinol 100	100	150	180	28	150	33/50
Flexinol 150	150	50	400	62	330	20/30
Flexinol 250	250	20	1,000	172	930	9/13
Flexinol 300	300	13	1,750	245	1,250	7/9
Flexinol 375	375	8	2,750	393	2,000	4/5

*Multiply by 0.0098 to get the force in Newtons (N).
†Cycles per minute, in still air at 20 °C.
‡LT = low temperature (70°C); HT = high temperature (90°C).

1. Create a "worm" or a "bug" that is moved by electronic muscles.
2. Try to create the device using solenoids.
3. Create a robot grip using SMAs or solenoids.
4. Try to design a mousetrap using SMAs or solenoids.
5. Build a feedback device using SMAs that allows the operator of a robot to sense in his hands when the robot runs into an obstacle.

7.6   Review Questions

1. What is an SMA?
2. What is the most common alloy used in SMAs?
3. Does the length of an SMA increase or decrease when it is heated?
4. What determines the force produced by a solenoid?
5. Where is the magnetic field of a solenoid more intense?
6. How can we determine the current across a solenoid when the resistance of the coil is given?

8

Stepper Motors

8.1 Purpose

Stepper motors may be used for locomotion, movement, positioning, and many other functions in which we require a precise control of the position of a shaft, lever, or a moving part of a mechatronic device. The purpose of this chapter is give to the reader basic information about the use of stepper motors in mechatronics and robotics projects, and to describe practical blocks and circuits.

8.2 Theory

The basic operation principle of a stepper motor is not much different from that of a dc or ac motor: they are formed by coils and magnets with a moving shaft that moves when power is applied. The difference is in the way the shaft is moved; they move the rotor by applying power to different coils in a predetermined sequence (stepped). Steppers are designed for fine control requirements and not only will spin on command but also travel any number of steps per second up to the maximum speed. Another stepper feature that cannot be matched by common motors is that stepper motors can hold their position and resist turning. Figure 8.1 shows the sym-

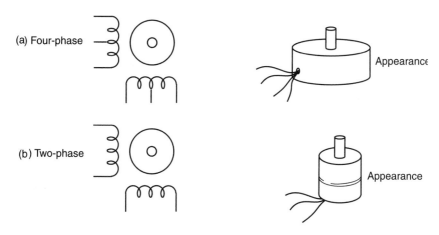

Figure 8.1 Stepper motors.

bol that has been adopted to represent a stepper motor and the appearance of the most common types.

The robot and mechatronic designer need not buy expensive stepper motors for these projects, since many of them can be found in good operating condition in old computer diskette drives, printers, and many other devices that have been retired.

8.2.1 How It Works

A stepper motor converts digital information into proportional mechanical movement. They are different from dc motors, which are controlled by changing the current flowing through them. Stepper motors are digital in operation.

Stepper motors can be found in three basic types: permanent magnet, variable reluctance, and hybrid. The way the windings are organized inside a motor determines how it works. The most common type is the four-phase stepper motor, but there are also two-phase and six-phase types. Figure 8.2 shows the most popular version: the four-phase stepper motor.

Inside this motor, we find four windings. Since each pair of windings has a common connection, this kind of motor can be easily identified by the six wires as shown in Figure 8.3.

In normal operation, the common wires are connected to the positive wire of the power supply, and the other wires are connected to ground for a short period of

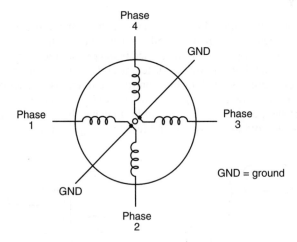

Figure 8.2   Four-phase stepper motor.

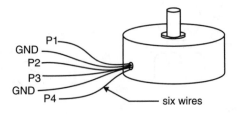

Figure 8.3   A four-phase stepper motor has six wires.

time—for as long you want to energize the correspondent winding. Each time the motor is energized, the motor shaft advances by a fraction of revolution. For the shaft to turn properly, the winding must be energized by a sequence of pulses or waves.

For instance, if you energize the windings A, B, C, and D in this sequence, the shaft turns clockwise. In the other hand, if you reverse the sequence, the motor turns counterclockwise. Figure 8.4 shows the sequence that is normally used to energize four-phase stepper motors.

Another way to energize a stepper motor is by applying an on-on/off-off sequence. This sequence is shown in Figure 8.5. It has the advantage of increasing the drive power of the motor and providing more precise shaft rotation.

Other common type of stepper motor is the two-phase unit, shown in Figure 8.6. This stepper motor is formed by two coils, as shown in the figure, and it can be easily identified by its four wires. This type of motor is energized using a different se-

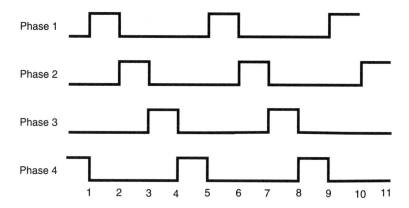

Figure 8.4   Sequence of pulses applied to a four-phase motor.

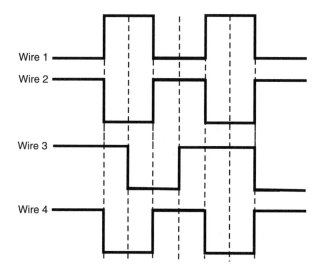

Figure 8.5   On-on/off-off sequence.

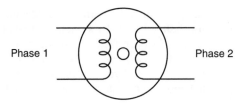

Figure 8.6   Two-phase stepper motor.

quence wherein the direction of the current across each widing is considered, as shown by Figure 8.7. In some types, the −V connection can be replaced by ground.

There are also stepper motors with more phases, e.g., the six-phase stepper motor, but they are not very common. Motors with more phases are more accurate but also more expensive.

For our purposes, embracing many applications in projects involving robotics and mechatronics, the four-phase stepper motor is recommended, and most of the blocks in this chapter are designed for this type of motor.

8.3   How to Use Stepper Motors

As we had seen, the windings of a stepper motor must be energized properly to achieve correct operation. This means that, when using a stepper motor, you'll need to know not only electrical specifications of the device but the mechanical specifications as well. The most important specs are described below.

8.3.1   Voltage and Current

Stepper motors are usually rated for 5, 6, or 12 V. Unlike with dc motors, overdriving the windings of a stepper motor is not recommended. Overvoltages of more than

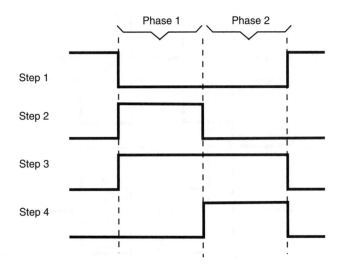

Figure 8.7   Phasing sequence for a two-phase motor.

30% of the rated voltage can burn the windings. The current ratings depend on the application (size and torque). Common types can draw currents in the range from 50 mA to more than 1 A. The higher the current and voltage, the higher the torque.

When designing a power supply for an application using a stepper motor, it is important to consider that the current ratings are given per winding. Therefore, the power supply must be able to supply at least twice the current per winding, or eight times the current per winding in a four-phase type.

8.3.2 Sequence

Although most stepper motors use one of the two sequences shown in the section on stepper theory, it is possible to find units that operate differently. When using such units, it is important to determine the correct operational pulse sequence.

8.3.3 Step Angle

When one pulse from the sequence is applied to the motor, it advances one step. This means that the shaft moves a specified number of degrees, referred to as *step angle*. The angle can vary among motor types in the range between 0.8 and 90°.

In a 90° stepper motor, four pulses move the shaft one complete turn as shown in Figure 8.8. However, it is more common to find stepper motors with step angles of 1.8°. This means that you have to apply 200 pulses to the control circuit to make the motor complete one revolution.

8.3.4 Pulse Rate

The pulse rate determines the speed of the motor. If you are using a 1.8° step angle motor, and you apply 200 pulses per second, this motor will run at 1 rotation per second or 60 rotations per minute (60 rpm).

Given the step angle, it is easy to calculate the rpm. The stepper motors are not intended for high-speed applications. The top recommended speed is in the range of 2 or 3 turns per second, or in the range of 120 to 180 rpm. It is important to remember that, in this kind of motor, the torque drops as the speed increases.

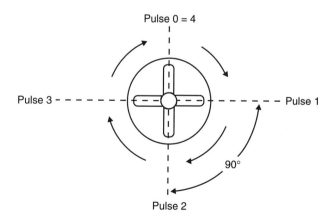

Figure 8.8   90° stepper motor.

8.3.5 Torque

The torque produced by a stepper motor is not high. A typical stepper motor can provide only a few grams per centimeter of torque in operation. This means that, in applications where high torque is needed, gearboxes must be added. Since the torque falls at higher speeds, this type of motor is best used in low-speed modes.

8.3.6 Braking Effect

If the current in a winding is maintained after a pulse is applied, the stepper motor cannot continue to turn. The shaft will be locked as if you applied a brake. A circuit that maintains the current in a winding to establish a fixed position effectively acts as electronic brake in a stepper motor.

8.4   Blocks Using Stepper Motors

When using a stepper motor, you need four types of blocks.

1. *Drivers.* These are the blocks that handle the high current needed by the windings of a stepper motor. They normally use medium- or high-power transistor for this task, or even ICs.
2. *Sequencers or translators.* This kind of block produces the sequence of pulses needed to drive the stepper motor. This circuit decodes the input pulses and translates them into the outputs needed for each type of motor.
3. *Steppers.* These blocks are used to produce the pulses at a rate that determines the speed of the step motor. They are normally in the form of oscillators.
4. *Controllers.* These blocks are formed by circuit that can act on the other blocks to brake, speed up, or slow down the motor, or even reverse its direction. These circuits can be ganged with sensor blocks and others block types.

Block 109   Standard Block Using Bipolar NPN Transistors

The block shown in Figure 8.9 is a driver for stepper motor handling up to 1 A. It is appropriate if the transistors are BD135/BD17 or BD139.

The sequence of pulses to position the motor is applied to the inputs P1 to P4 for an appropriate translator (see the following blocks). The circuit needs about 12 mA per input to drive a 500 mA motor, making it compatible with TTL or CMOS logic.

The transistors can be replaced with powerful units (see Chapter 3 for more information). If low-gain transistors are used, you'll probably need more current in the input. The next block, using Darlington transistors, can be important if high-power stepper motors are used. The transistors must be mounted on heat sinks.

Block 110   Standard Block Using Darlington NPN Transistors

The block shown in Figure 8.10 needs less power to drive high-current stepper motors from TTL, CMOS, or other drive circuits. The circuit requires less than 1 mA to drive motors, depending on the transistors. For the recommended transistor, currents up to 1 A can be controlled. If you need more current, see Chapter 3 for directions on how to select a more powerful transistor.

A translator must be used to produce the sequence of pulses according to the position or the speed needed in your applications. Remember that the power transistors must be mounted on heat sinks.

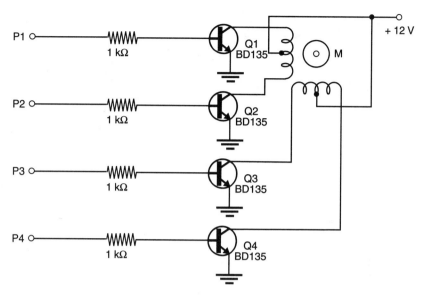

Figure 8.9   Block 109: standard circuit using bipolar transistors.

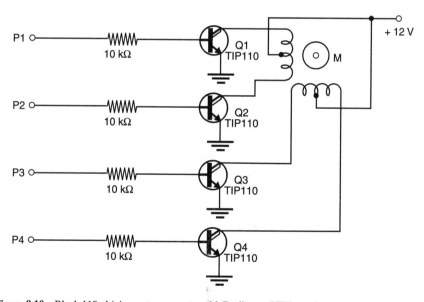

Figure 8.10   Block 110: driving a stepper motor with Darlington NPN transistors.

Block 111   Driving a Stepper Motor with PNP Transistors

Depending on the application, the stepper motor can be driven from negative logic (inverted pulses). Instead of adding logic inverters, you can use this block, which operates with negative logic (see Figure 8.11).

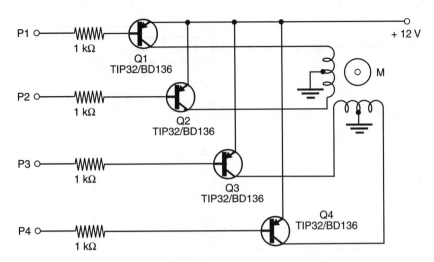

Figure 8.11   Block 111: stepper motor driver using PNP transistors.

Each winding is on when the input of the corresponding transistor is set to the low logic level. The circuit must be connected to a sequencer or other circuit to give the desired sequence of pulses to make the motor run or to put it in a predetermined position. Using the recommended transistors, the block can drive motors with up to 1 A from TTL or CMOS logic. See Chapter 3 if you want to use more powerful transistors.

Block 112   Using PNP Darlington Transistors

More power can be provided to stepper motors using the block suggested by Figure 8.12. This block is the Darlington version of the previous block, where each transis-

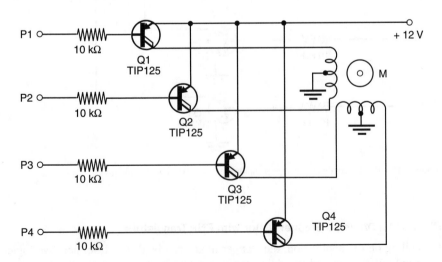

Figure 8.12   Block 112: using PNP Darlington transistors.

tor is on when the input (base) is set to the low logic level. The recommended transistors need less than 1 mA to drive motors up to 1 A. If you want to drive more powerful motors, see Chapter 3 for information about other Darlington transistors suitable for the task.

Block 113   Driving a Stepper Motor with Power MOSFETs

Power MOSFETs can be used to drive stepper motors as shown by the block in Figure 8.13. The high-impedance input characteristic of power MOSFETs makes them ideal for this purpose. Power MOSFETs can drive motors with currents of many amperes, depending on the type. This circuit can be driven from TTL and CMOS logic and other blocks.

Block 114   Step Generator Using the 555 IC

The circuit shown in Figure 8.14 produces the pulses necessary to drive a translator coupled to a stepper motor. The circuit is an oscillator wherein the output frequency determines the speed of the stepper motor when it operates as a common motor.

S1 is a switch used to step down the frequency range when closed. The circuit can be powered from supplies in the range between 5 and 18 V. P1 adjusts the frequency, and the output is compatible with TTL and CMOS logic.

P1 can be replaced by a resistive sensor such as an LDR to control the speed from external signals (see Section 10 for more details).

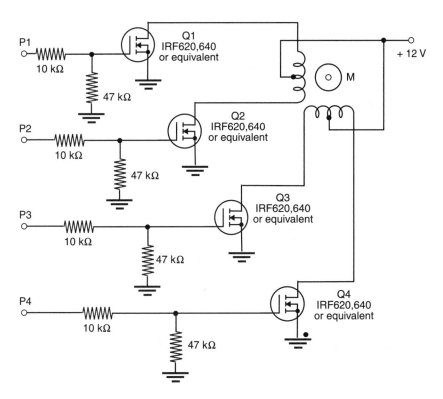

Figure 8.13   Block 113: using power MOSFETs.

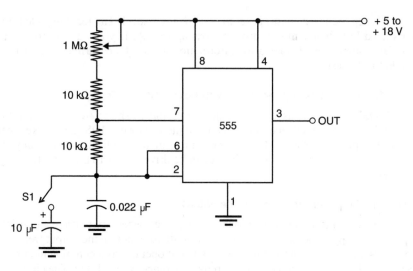

Figure 8.14 Block 114: step generator using the 555 IC.

Block 115   Step Generator Using the 4093 IC

Another step generator is shown in Figure 8.15. This circuit uses a CMOS IC and can drive translators of the same logic type.

It is also an oscillator that produces the necessary pulses at a frequency that determines the speed of the stepper motor. When S1 is closed, C1 is put in the circuit, stepping down the frequency and the speed of the motor. P1 adjusts the speed range of the motor. This component, as in the previous block, can be replaced with a resistive sensor (see Chapter 10 for more details).

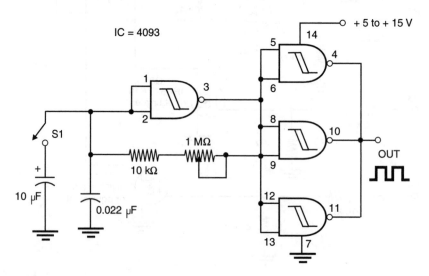

Figure 8.15 Block 115: step generator using the 4093 IC.

Block 116   Step-by-Step Generator

Stepper motors can be controlled directly from a switch placed at the input of a translator. As each pulse is produced when the switch is closed, the motor will advance one step.

The main problem with this type of circuit is that the switch must be *debounced*. Any voltage spike produced when the switch is closed and opened can be interpreted by the translator as an additional pulse. Therefore, when closing and opening the switch to produce only one pulse, the circuit can receive many pulses and advance the motor more than one step. (See Chapter 9 for debounce circuits.) The block shown in Figure 8.16 is a stepper that can be used to produced only one pulse when a sensor or a switch is momentarily pressed.

The main advantage of this circuit is that the duration of the pulse depends only on R2 and C1, and not from the time the switch (or sensor) is kept on. The circuit can drive TTL and CMOS blocks and be powered from voltages in the range between 5 and 15 V.

Block 117   Control for Two-Phase Stepper Motor

A two-phase stepper motor can be controlled easily using the circuit shown in Figure 8.17. The transistors are chosen according to the current requirements of the motor. Chapter 3 will help the reader find the correct transistor based on the relays if these components need more than 50 mA to be driven. This block must be driven by a translator as in the case of four-phase motors.

Note the sequence necessary to drive them in Section 8.2, Theory. We also need to observe that, in some two-phase motors, the –6V can be replaced by ground.

8.4.1   Using ICs

The robotics and mechatronics designer can locate many ICs designed for stepper motor control. These ICs can start from simple units that use only four transistors to drive the motors, to more complex units that include the steppers, translators, and many other functions. The following blocks show how some of these ICs can be used in our projects.

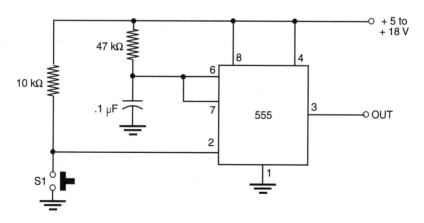

Figure 8.16   Block 116: step-by-step generator.

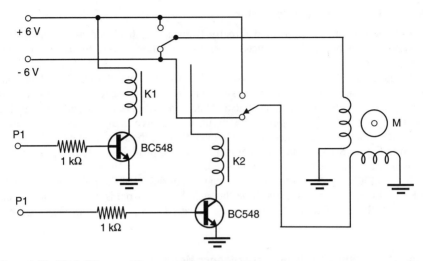

Figure 8.17   Block 117: control for a two-phase stepper motor.

Block 118   Driving a Stepper Motor with the ULN2002 and ULN2003

Stepper motors with current requirements up to 500 mA can be driven directly from ICs of the ULN200X series. These ICs are formed by paired power transistors, including the base resistor and the diode to protect them against the voltage spikes generated when the stepper motor windings are switched on and off. The difference between the two types is that resistors of different values are placed at their inputs.

In the ULN2003, the we find a 2,700 Ω resistor, matching the characteristics of the IC with TTL logic. In the ULN2004, the resistors at the input of each stage are 10.5 kΩ units, matching its characteristics with CMOS logic.

The block in Figure 8.18 shows how the ULN2002/3 can be used to drive a stepper motor using up to 500 mA. This IC doesn't need a heat sink and can be powered with voltages up to 12 V.

Block 119   Driving Stepper Motors with the MC1413/MC1416

The MC1413 and MC1416 ICs, from Motorola, are equivalent to the ULC2003 and can be used to drive stepper motors as shown in Figure 8.19. The MC1413 has an equivalent circuit for each stage, as shown in Figure 8.20a, and the MC1416 has the configuration shown in Figure 8.20b.

The difference is that the MC1413 has characteristics that match it with TTL logic, and the MC1416 with CMOS logic with voltages from 8 to 18 V. The block can control currents up to 500 mA per winding, and the maximum voltage in the motor is 50 V.

Block 120   Complete Stepper Motor Control with the SAA1027

The SAA1027 IC is designed to include all the functions needed to control a stepper motor with currents up to 350 mA. The SDA1027 is manufactured by Signets, and the basic application is shown by Figure 8.21.

The bias resistor must be 470 Ω × 1/2 W if the stepper motor needs less than 250 mA current. If the current is in the range between 200 and 350 mA per phase, this resistor value will be reduced to 150 Ω × 1 W.

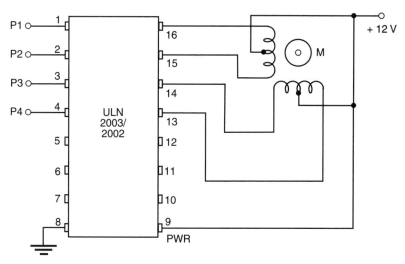

Figure 8.18   Block 118: stepper motor control using the ULN4202.

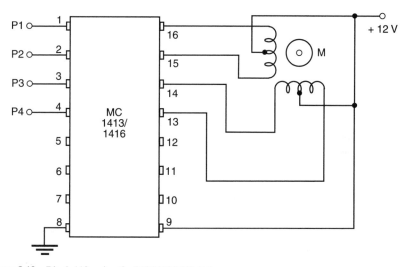

Figure 8.19   Block 119: using the MC1413/MC1416.

This block has three inputs:

1. The pulses that move the motor are applied to the trigger input.
2. To the direction input, we apply the logic level that determines whether the stepper motor runs forward or backward.
3. The set input, when connected to ground, continuously supplies power to two of the windings, braking the motor.

Stepper motors with higher current ratings can be driven using one of the driver blocks shown at the beginning of this chapter.

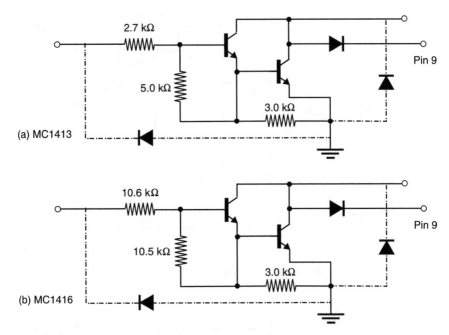

Figure 8.20   Block 119: driving stepper motors with the MC1413/MC1416 IC.

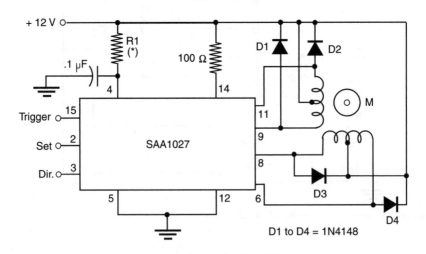

Figure 8.21   Block 120: stepper motor control using the SAA1027.

Block 121   Complete Stepper Motor Control Using the UCN4202

The UCN4202 (Sprague) is another IC that contains all the functions required by a complete stepper motor control. This IC is used in a basic configuration as shown by Figure 8.22.

Stepper motors with up to 500 mA per winding can be controlled by this circuit, without the aid of any driver transistor stage. If the current requirements of the mo-

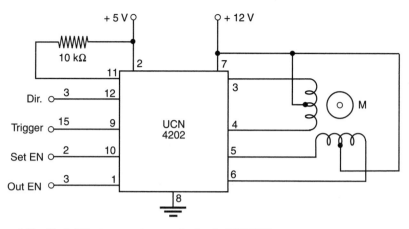

Figure 8.22   Block 121: stepper motor control using the UCN4202.

tors are higher, you can add one of the driver blocks shown at the beginning of this chapter.

This IC's control inputs are as follows:

- The *direction* determines if the motor runs forward or backward, according to the logic level.
- The *trigger* enable is where the control pulses are applied, determining the speed of the motor.
- In the *step* enable, you can brake the motor, as in the *output* enable.

Block 122   LED Monitor for Stepper Motor Operation

The circuit shown in Figure 8.23 can be used to monitor the operation of each phase of a stepper motor. The LED corresponding to each phase will glow when it is acti-

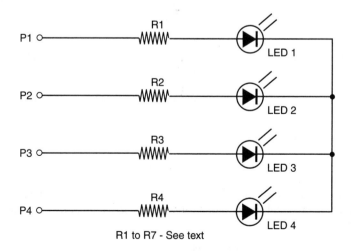

Figure 8.23   Block 122: LED monitor.

vated. The resistors' value depends on the power supply voltage according to the following table:

Voltage	R1/R2/R3/R4
5 V	330 Ω
6 V	470 Ω
9 V	680 Ω
12 V	1 kΩ
15 V	1.2 kΩ
18 V	1.8 kΩ
24 V	2.2 kΩ

The resistors are 1/8 W units. LEDs of any color can be used in this indicator.

8.5   Additional Information

Stepper motors can be found in many sizes, voltages, and currents (given a particular output power). Table 8.1 gives information about other transistor arrays or power stages that are recommended to drive stepper motors. Figure 8.24 shows the pinout of these ICs.

Table 8.1   Other Recommended ICs

Type	Current	No. of Stages	Manufacturer	Observations
ULN2803	500 mA	7		Equivalent to the UCN2003 but in 18-pin package
MC1413	500 mA	7	Motorola	TTL compatible
MC1414	500 mA	7	Motorola	CMOS compatible
SN75465	500 mA	7	Texas Instruments	1050 Ω series resistor at each input, TTL compatible
SN75466	500 mA	7	Texas Instruments	2700 Ω series resistor at each input, CMOS compatible
SN75468	500 mA	7	Texas Instruments	Indicated for CMOS and TTL with 5 V supply
SN75469	500 mA	7	Texas Instruments	10,500 Ω series resistor at each input, CMOS with 6 to 15 V supplies

8.6   Suggestion for Projects

The blocks shown in this chapter are not the only ones that can be ganged to produce interesting projects in mechatronics and robotics. The reader can use his imagination and gang together blocks shown in other chapters (including the next one). Some suggestions for practical (or theoretical) work on the circuits presented in this book are as follows:

1. Create a robot in which the wheels are controlled by stepper motors, and the direction is controlled by the steps.

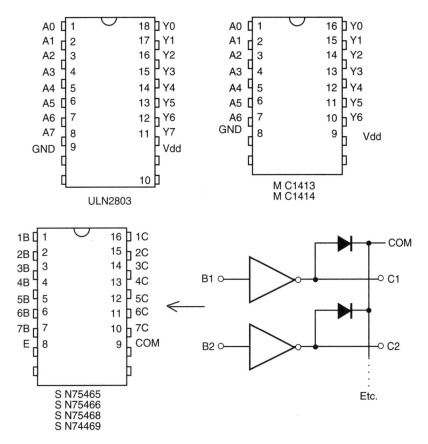

Figure 8.24   Pinout of ICs for stepper motor control.

2. Design an arm in which movements are controlled by potentiometers and keys using stepper motors.
3. Design a simple direction control using stepper motors.
4. Design a plotter that will move a pen to any position on a piece of paper using only two stepper motors.
5. Design a clock with two stepper motors in which one move the minute hand, and the other the hour hand.

8.7   Review Questions

1. What is the basic difference between a stepper motor and a dc motor?
2. Can we use a stepper motor for the same purposes as a dc motor?
3. Can a stepper motor be controlled by digital circuits?
4. How many coils does a two-phase stepper motor use?
5. What is a translator?
6. What is the braking effect?
7. What determines the speed of a stepper motor?
8. How can we reverse the direction of a stepper motor?

9

On-Off Sensors

9.1 Purpose

The purpose of this chapter is teach to the reader how mechanical switches, reed switches, and home-made devices can be used as sensors. Pairs of metal blades, common on-off switches, pushbuttons, microswitches, and reed switches can be used as sensors in robotics and mechatronics projects. In this chapter, we include blocks using these devices to add some important features such as debouncing, inverting, and delaying.

9.2 Theory

We often need to turn a circuit on or off when an event occurs; this is a basic function in robotics and mechatronics projects. A motor can be reversed when the robot is stopped by an obstacle or when a mechanical arm strikes an object. A mobile robot can change its direction if a sensor detects a wall in its way.

Starting with some common components and configurations, you can add mechanical sensors to your projects. Several are described below.

On-off switches—microswitches. On-off switches can easily be adapted for use as position sensors in many applications. Figure 9.1 shows how can we adapt a slide on-off switch (SPST) to be used as an impact detector.

This switch can be placed in front of a robot and used to detect when it runs into any obstacle. Soft-touch types are ideal for these applications, since it does not take much power to close them.

Microswitches, e.g., those used in industrial machines, are also ideal for applications involving position sensors such as the ones shown by Figure 9.2. These

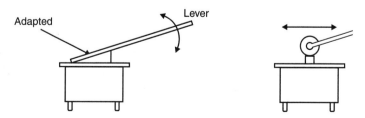

Figure 9.1 Using SPST switches as sensors.

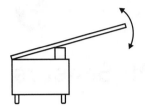

Figure 9.2   Microswitch used as a sensor.

switches are small and sensitive, needing only a small force to be engaged. Many types of mechanisms can be adapted to common switches to convert them into sensors. Levers can increase the sensitivity of a microswitch. Pushbuttons can also be used as sensors, as shown by Figure 9.3. It is important to choose one that can be easily closed by the mechanism to which it is attached.

Reed switches.   Reed switches are formed by a glass enclosure that is filled with an inert gas, inside of which are blades with electric contacts. This is shown in Figure 9.4. The presence of a magnetic field acts on the blades, making them to touch each other and thereby closing the circuit. Reed switches are very sensitive and can be used as sensors as shown in Figure 9.5.

In operation, we use the field of a magnet to engage the switch. We simply install the magnet in the mobile device or in any mechanical part whose movement or position change must be detected.

Home-made sensors.   Two or three blades placed together as shown in Figure 9.6 form a mechanical sensor that has many applications in robotics and mechatronics. The blades and the contacts can be found in old relays and switches. The reader only needs to exercise some care to make sure that the contacts are in good working condition. Darkened or deformed contacts indicate burns that can reduce the efficiency of the sensor. If present, it is best to discard the component and find a better one.

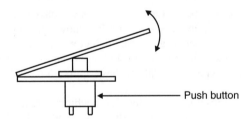

Figure 9.3   Pushbutton used as a sensor.

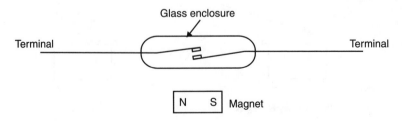

Figure 9.4   A reed switch.

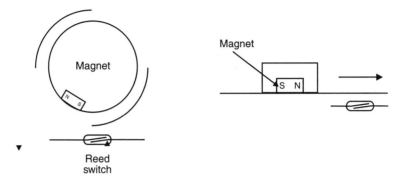

Figure 9.5   Using reed switches as sensors.

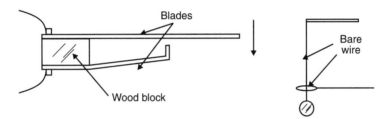

Figure 9.6   Home-made sensor.

The blades can be activated in many ways. The figure shows some suggested mechanisms that can be used to activate this kind of sensor. A long spindle, for instance, can be used to create a position or obstacle sensor.

9.2.1   Debouncing

When a switch is closed, electric contact isn't established immediately. Mechanical contacts such as found in switches, relays, and sensors produce noise during their *settling time.* They bounce and make repeated contact during the first few milliseconds of action as shown by Figure 9.7.

Logic circuits such as TTL and CMOS, and even some other electronic circuits, are fast enough to recognize each and every bounce (noise pulse) as a separate impulse or logic level change. You close the switch one time, expecting to add one

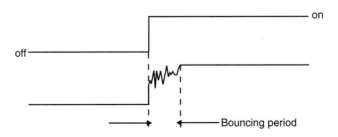

Figure 9.7   Contact bounce when turning on a switch.

count to the command chain, but the circuit will jump many commands. To avoid this problem when using switches, reed switches, relays, or other on-off sensors with mechanical contacts, a proper conditioning (debouncing) circuit must be used.

The simplest circuit is formed by an RC network as shown in Figure 9.8. The voltage across the capacitor rises gently, even when many pulses are produced by the switch.

Other circuits, as we will show in the following blocks, can help the robotics and mechatronics designer to avoid problems created by bouncing contacts in sensors, switches, and other devices that use mechanical contacts.

9.2.2 Switches as Sensors

Many types of switches can be used as sensors or adapted to that task. Momentary switches can be used as bump sensors, navigation feelers, and limit sensors. The most useful are microswitches and pushbutton switches such as the ones shown in Figure 9.9.

9.2.3 Reed Switches

A reed switch is formed by a glass bulb filled with an inert gas. Inside the bulb, we find two or more metal blades as shown in Figure 9.10. When a magnetic field acts on the blades, they bend and touch one another, closing the external circuit.

Reed switches are very sensitive but can't operate with large currents. Typical currents are in the range between 50 and 500 mA. Since they can switch a circuit very quickly, they can be used in high-speed applications, sensing the rotation of a shaft or any mobile part of a mechanism as shown in Figure 9.11.

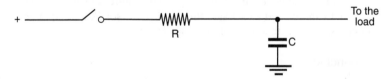

Figure 9.8 Simple debouncing.

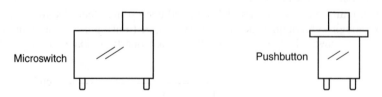

Figure 9.9 Switches suitable for mechanical sensors.

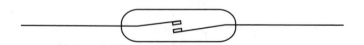

Figure 9.10 The reed switch.

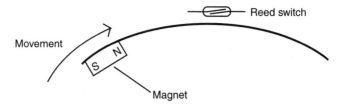

Figure 9.11   Sensing rotation with a reed switch.

Attaching a magnet to a wheel and placing a reed switch in an appropriate position, it is possible to produce a pulse every revolution. The pulses can be used to control the speed of the wheel or to excite a tachometer.

9.2.4   Home-Made Sensors

Many home-made sensors can be designed by the robotics and mechatronics builder. One of them is the inclination sensor shown in Figure 9.12.

The same configuration, using a metal wire and a hoop, can be used as an aceleration sensor. Since these sensor can produce only short bursts of current when activated, appropriate circuits can be used as signal conditioners.

9.2.5   Programmed or Sequential Mechanical Sensors

Simple arrangements using wheels and screws can produce position sensors or programmed sensors such as the ones shown in Figure 9.13. In the first case, a cylinder covered by a metal plate can be used as a position sensor. Blades in the array touch the metal only at the points where no isolation exists. The designer can design the covered and uncovered positions of the cylinder according to the signal to be produced by the switches. This home-made sensor can be used to convert the position of an arm coupled to a cylinder into digital information.

9.3   Basic Blocks Using Sensors

The following are blocks that use an on-off sensor such as described in the previous paragraphs. Many blocks that we have already discussed can be used with these sensors, as appropriate for particular applications. Some descriptions of these blocks are provided in the following discussion.

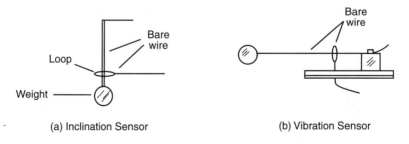

(a) Inclination Sensor   (b) Vibration Sensor

Figure 9.12   Home-made sensors.

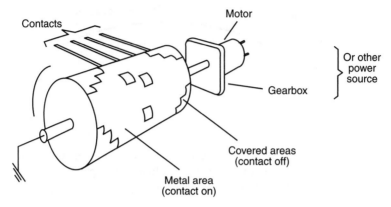

Figure 9.13   Sequential sensor.

Block 123   Turning a Load On

The simplest way to use a home-made, normally open (NO) mechanical sensor is shown in the Figure 9.14. The load (motor, lamp, circuit, solenoid, etc.) is placed in series with the sensor. The designer just needs to be careful to keep the load below the maximum current capability of the sensor contacts.

Block 124   Turning a Load Off

The simple block shown by Figure 9.15 shows how a normally closed (NC) sensor is used to control a load. The recommendations are the same as in the previous block. Note that the sensor keeps the contacts closed (or open) only while a force acts on it.

Block 125   Contact Conditioner with a Capacitor

If the load is sensitive to contact bounce that occurs when closing or opening the switch, a contact conditioner must be used. The simplest contact conditioner is the

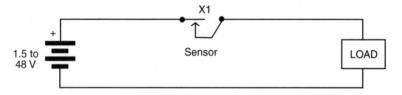

Figure 9.14   Block 123: turning a load on.

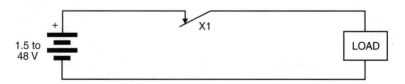

Figure 9.15   Block 124: turning a load off.

one shown in Figure 9.16, which uses a capacitor. The value of the capacitor depends on the load. For inductive loads, electrolytic capacitors in the range of 10 to 1,000 µF are recommended. For other loads, capacitors starting at 0.1 µF can be used.

Block 126   Low-Current Turn-On Sensor with Contact Conditioner

Reed switches can't be used to control large currents. The circuit shown in Figure 9.17 is recommended for low-current sensors, including those using reed switches.

Resistor R can assume values between 1 and 10 kΩ, depending on the transistor. See Section 9.3.1 to determine which blocks described in Chapter 3 can be added to this block to control high-power loads. The contact conditioner capacitor must assume values between 0.05 and 1 µF, depending on the application.

In the same block, we see how can the load be replaced with a resistor (1 kΩ) to produce a signal for an external circuit. If a BC548 is used in this circuit, resistor R must be a 10 kΩ component and C a 0.047 µF capacitor. Powering the circuit with 5 to 12 V, the circuit can be used to drive TTL or CMOS inputs from mechanical contact sensors. The output of this circuit goes to the low logic level (0) when the sensor is closed. See Section 9.3.1 for blocks that can be controlled by this circuit.

Block 127   Low-Current Turn-Off Sensor with Contact Conditioner

The block shown by Figure 9.18 turns a load off when the sensor is closed. The transistor is chosen according to the load current (see Chapter 3 for more information). The capacitor can have values between 0.01 and 10 µF, depending on the application. As in the previous block, the load can be replaced with an 1 kΩ resistor. With R = 10 kΩ and C = 0.047 µF, the circuit can be used as a contact conditioner driving

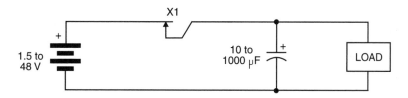

Figure 9.16   Block 125: debouncing with a capacitor.

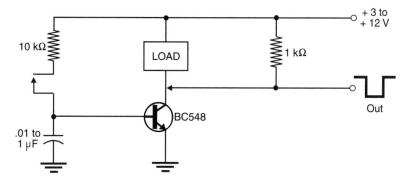

Figure 9.17   Block 126: low-current sensor.

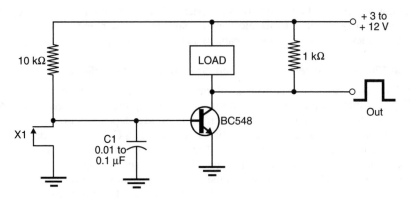

Figure 9.18 Block 127: turn-off sensor with contact conditioner.

TTL and CMOS circuits. See Section 9.3.1 to determine which blocks can be added to this circuit to control high-current loads.

Block 128   Contact Conditioner Using the 555 IC

The best contact conditioner for applications using logic circuits (TTL or CMOS) are the ones based in monostable configurations. The monostable configuration shown by Figure 9.19 uses the 555 IC and is ideal for this purpose.

This circuit produces an output pulse with a duration determined by R and C. With a 10 kΩ resistor and 0.1 µF capacitor, individual short-duration pulses can be generated from mechanical sensors. The same circuit can be used in other applications such as a long-period monostable or a timer, as we will see in other parts of this book. Section 9.3.1 indicates which circuits can be used to increase the current capabilities of this circuit.

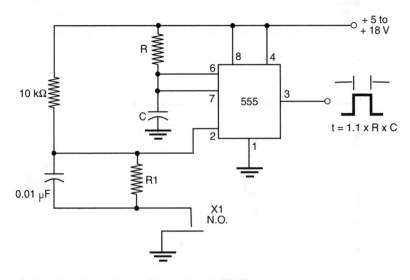

Figure 9.19 Block 128: contact conditioner using the 555 IC.

Block 129   Contact Conditioner Using the 4093 CMOS IC (I)

The contact conditioner uses one of the four Schmitt AND gates existing in a 4093 CMOS IC (Figure 9.20). The capacitor can have values in the range between 0.01 and 0.1 μF. Resistor R1 must have values between 47 kΩ and 1 MΩ. Its function is to discharge the capacitor quickly when the sensor is open and waiting for a new pulse. If the capacitor is charged when the sensor is closed again, the gate rests with the output low and does not change this state.

Note that this circuit passes from the high logic level to the low logic level when the sensor closes its contacts. The circuit is recommended as a contact conditioner for driving CMOS logic. See Section 9.3.1.

Block 130   Contact Conditioner Using the 4093 CMOS IC (II)

The difference between this circuit and the one shown in the previous block is that, in this circuit, the output goes to the high logic level when the sensor is closed (Figure 9.21).

The components (capacitor and resistor) are chosen according to the bounce characteristics of the sensor and are in the same range of values described in the last block. See also Section 9.3.1. In it you'll find blocks that can be added to the output of this circuit to increase its power.

Block 131   TTL Contact Conditioner Using the 7400 IC

The block shown by Figure 9.22 uses a TTL IC and must be powered from a 5 V power supply. When the sensor is closed, the output of the logic block goes to the

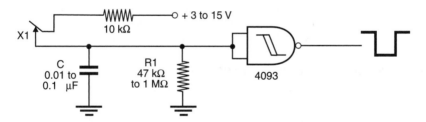

Figure 9.20   Block 129: contact conditioner using CMOS IC.

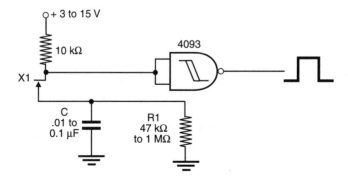

Figure 9.21   Block 130: contact conditioner II.

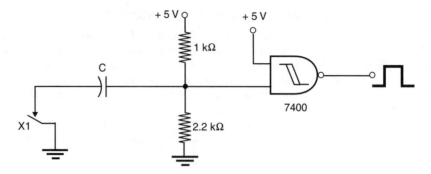

Figure 9.22   Block 131: using the 7400.

high logic level and can drive a power stage such as shown in previous parts of this book. The capacitor determines the time constant of the circuit, and typical values to eliminate bouncing problems are in the range between 0.047 and 0.47 µF. See also Section 9.3.1.

Block 132   Contact Conditioner for Two Sensors—Bistable

The circuit shown in Figure 9.23 is recommended for applications in which two mechanical sensors are used. The circuit turns on with the output passing from the low logic level to the high logic level when S1 is closed. Then the circuit latches and passes again to the low logic level in the output if S2 is closed momentarily.

Since it is a TTL circuit, the power supply must be a 5 V unit. Any equivalent of the 7404, or any inverting function, can be used in this block. Section 9.3.1 suggests blocks that can be controlled by this circuit, increasing the output power.

Block 133   Contact Conditioner for SPDT Sensor

Two contact sensors operating as a single pole, double throw (SPDT) switch can be conditioned by the circuit shown in Figure 9.24. This circuit uses two NAND gates of a 7400 TTL IC. The position of the contacts determines whether the output of the circuit goes to the high or low logic level.

If the circuit is intended to drive high-power loads, a transistorized stage such as the ones described in Chapter 3 must be used. See Section 9.3.1 for more information.

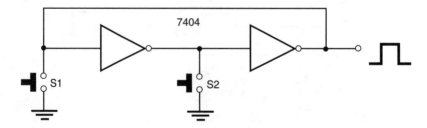

Figure 9.23   Block 132: contact conditioner for two sensors.

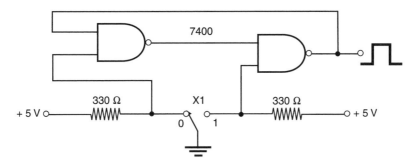

Figure 9.24   Block 133: TTL circuit for an SPDT sensor.

Block 134   Contact Conditioner for Two Sensors

Another block that can be used as a contact conditioner is shown in Figure 9.25. This block also uses TTL ICs and must be powered from a 5 V power supply. The logic level of the inputs determines the output logic level of the circuit. See Section 9.3.1 for blocks that can be driven by this circuit.

9.3.1   Notes on Compatible Blocks

The following blocks can be added to the output of the previously described contact conditioners to control high-power loads. It is up to the designer decide which is the correct block for a particular application, taking into account the electrical characteristics and the desired performance.

- Blocks 30 through 37: compatible with transistors (bipolar, CMOS, TTL, CMOS logic and the 555 IC).
- Blocks 54 and 59: controlling dc motors (H-bridges)

9.3.2   Controlling Motors and Loads

The direct control of dc motors and other loads using mechanical sensors is described in the following blocks. In some cases, capacitors must be added in parallel with the loads to avoid bounce problems or erratic operation.

Block 135   Direction Control Using a Mechanical Sensor

Figure 9.26 shows how to connect a SPDT sensor to a motor for direction control. Observe that this circuit has the same function as an H-bridge. The sensor can be

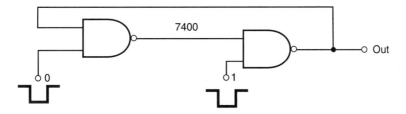

Figure 9.25   Block 134: contact conditioner for two sensors (TTL)

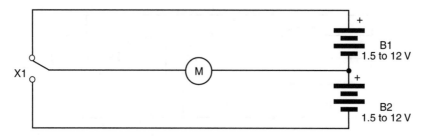

Figure 9.26   Block 135: direction control using a mechanical sensor.

used to control many of the bridges described in previous parts of this book. Remember that the current-handling capabilities of the sensor must be higher than the current requirements of the controlled motor.

Block 136   Controlling Two Loads

Figure 9.27 shows how an SPDT sensor can be used to control two loads. The loads can be relays, solenoids, motors, or any other circuit. For more information, review previous blocks describing how SPST switches can be used to control many functions of motors, relays, and other loads.

Block 137   High-Low Sensor

Figure 9.28 shows how an SPDT sensor can be used to send a bit of data to a circuit. This bit can be used to control a stepper motor, a dc motor, or any other load by a logic circuit.

Block 138   High-Power Motor Control

Figure 9.29 shows how the SPDT sensor can be used to control the direction of a dc motor with a dual-voltage power supply. In blocks with motor controls, the reader can obtain a wider selection of configurations by using transistors in the controls.

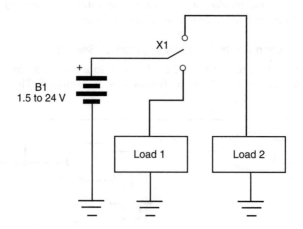

Figure 9.27   Block 136: controlling two loads.

Figure 9.28   Block 137: high-low sensor.

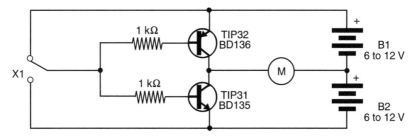

Figure 9.29   Block 138: high-power control.

The transistors can be the BD135/136 pair for motors up to 500 mA. For more powerful motors use the TIP31/32 pair. See Chapter 3 for more information about the
transistors.

Block 139   Timed Sensor Using a 555 IC

When the sensor sets the input of this block to the low logic level, its output goes to
the high logic level, activating the relay. The relay will remain on for a time interval
determined by the time constant RC.

The block shown in Figure 9.30 can be used as a timed sensor, activating a load at
time intervals ranging from a few seconds to more than 30 min, depending on the

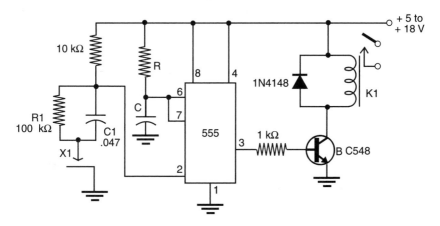

Figure 9.30   Block 139: timed sensor (555 + NPN).

components used in the RC network. For a 1,000 µF capacitor and a 1 MΩ resistor, the time interval can be extended to up to 15 min.

The output transistor in this block is an NPN BC548, meaning that the relay is on when the 555 is set with the output in the high logic level. You can use other transistors, as described in Chapter 3, to drive high-current loads directly. The circuit can be powered from 5 to 12 V supplies.

Block 140   High-Current Control for Mechanical Sensors

The circuit shown in Figure 9.31 can be used to control loads up to 3 A from such low-current sensors as reed switches or home-made switches. It is a momentary control, since the load is on only during the time the sensor is open.

When the sensor is closed, we find a 1.25 V voltage in the output of the circuit. When the sensor is open, the voltage in the output rises to a value adjusted by P1. The LM350T can supply voltages up to 25 V. The input voltage must be at least 3 V higher than that desired in the output.

This circuit is recommended for applications where, in the low state (load off), the voltage can be as low as 1.25 V). If a zero-volt output is desired, a negative voltage of 1.25 V must be applied to the sensor instead to shunt it to ground. The LM350 must be mounted on a heat sink.

Block 141   Multi-voltage Control for Mechanical Sensors

Many sensors (reed switches, home-made sensors, etc.) can be used to generate different voltages in a load. The block illustrated in Figure 9.32 shows how this can be done using a single LM350 IC. This IC can control up to 3 A in a voltage range from several volts to 25 V, as in the previous block.

When all the sensors are open, the voltage found in the output of the block (across the load) is determined by Px. This control can adjust the output for voltages between 1.25 and 25 V.

When any sensor is closed, the voltage changes to a value determined by the adjustment of the corresponding variable resistor. For instance, when X1 is closed, the output voltage falls to a value determined by the value of P1 and Px in parallel.

The LM350 must be mounted on a heat sink, and the input voltage must be at least 3 V higher than the maximum programmed voltage in the output. This block can be controlled by a drum position sensor as described in Section 9.2, in theory.

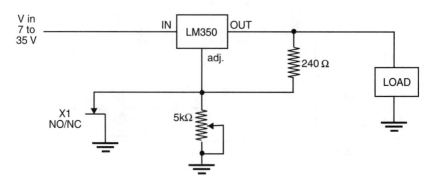

Figure 9.31   Block 140: controlling loads up to 3A.

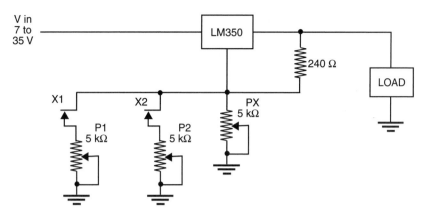

Figure 9.32   Block 141: multi-voltage control.

Many blades can make contact with a drum (cylinder) where the voltages are programmed according to the position.

Block 142   Priority Switch

An important block, with many applications in robotics and mechatronics, is the priority switch. This circuit can detect which sensor (or switch) was activated first and turn on a circuit to disable the other.

Our first block of a priority switch is shown in Figure 9.33 and uses SCRs as basic elements. Assuming that both SCRs are off and S1 is closed before S2, the SCRs are activated first, providing current to Load2.

With the SCR2 turned on, the voltage in its anode falls to about 2 V, which is not enough to turn on SCR1 if S2 is closed.

Note that this circuit is self-latched, meaning that to turn the triggered SCR on or off, the power must be cut momentarily. The SCRs must be mounted on heat sinks if the loads draw more than 500 mA.

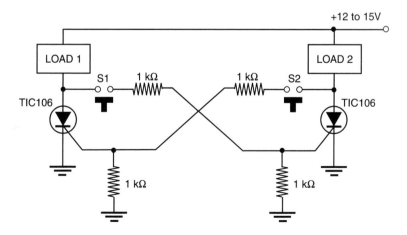

Figure 9.33   Block 142: priority switch.

Block 143   Tachometric Sensor

The turn-on/off speed of a sensor can be used to measure how fast a robot walks or runs or how fast a mechanism moves from one position to other. The circuit shown in Figure 9.34 converts the on/off frequency of a sensor into a proportional voltage.

This circuit can be used to sense the speed of any mechanism, coupling to it a magnet to act on a reed switch or using it to control mechanical contacts. The time constant of the RC network is calculated to give the desired range of voltages in the output as a function of the frequency of the input pulses generated by the sensor. The circuit can be powered from 5 to 18 V supplies.

Block 144   Missing Pulse Detector

An important function for projects in robotics and mechatronics is the *missing pulse detector.* This circuit keeps its output in the high logic level, presenting the power supply voltage while a constant pulse sequence is applied to the input as shown in Figure 9.35.

If one or more pulses are missing in this sequence, the output of the circuit passes for a brief time interval to the low level, generating a "flag" signal. This signal can be used to trigger an alarm or another circuit.

The block is shown in Figure 9.36 and is a monostable based in the 555 IC. R and C are calculated to provide the circuit with a time constant that is longer than the distance between two incoming pulses but shorter than the distance between two

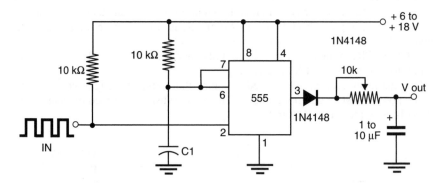

Figure 9.34   Block 143: tachometric sensor.

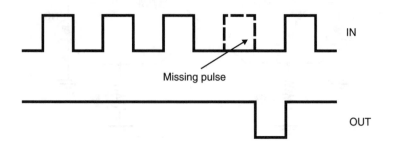

Figure 9.35   The missing pulse detector.

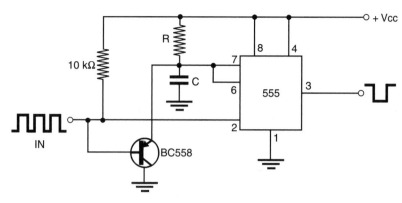

Figure 9.36   Block 144: missing pulse detector.

pulses. This means that the circuit will be retriggered by the next pulse before the end of the timing process, keeping the output high. If a pulse is missing, the time constant ends before the next pulse has arrived, and the output goes to the low level in an instant, producing the flag output signal.

The time constant is given by $t = 1.1 \times R \times C$, and R can be kept in the range between 1 kΩ and 1 MΩ.

The bipolar 555 can drain about 200 mA, but we also can build a CMOS version (TLC7555) that can drain and supply 200 mA. If the load needs more current, an output stage must be used such as the ones described in Chapter 3.

9.4   Suggested Projects

1. Create a bump-and-obstacle sensor system using home-made sensors and a circuit that can control all the movements of this robot.
2. Design a mechanical arm with on-off sensors and the control circuit for it.
3. Create a system that can sense the incline of the place where the robot walks (or runs) and compensate for it by adjusting the speed/power of the motors. Use the sensors described in this chapter.

9.5   Additional Information

TTL gates and logic inverters that can be used in contact conditioners and other logic applications are shown in the following table:

Type	Function
7400	4 2-input NAND gate
7402	4 2-input NOR gate
7404	Hex inverter
7414	Hex Schmitt triggers—inverting
7486	4 exclusive-or gate

Figure 9.37 shows the packages and pinouts for these ICs. Remember: any regular TTL output can sink 16 mA but source only 0.4 mA.

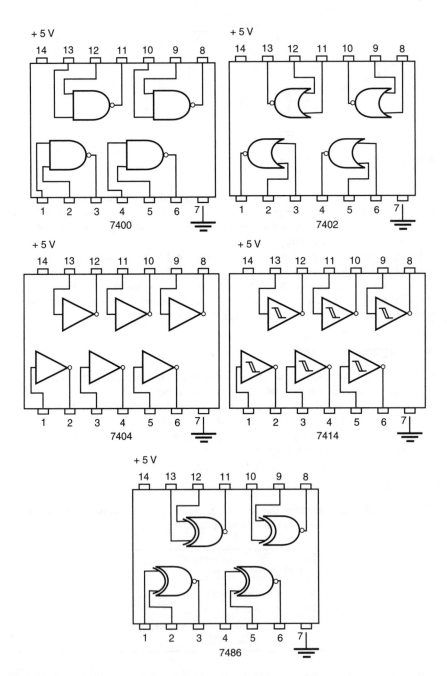

Figure 9.37   TTL ICs.

Useful CMOS gates are listed in the following table:

Type	Function
4001	4 2-input NOR gate
4011	4 2-input NAND gate
4009	Hex buffers (inverting)
4010	Hex buffers (non inverting)
4093	4 2-input NAND Schmitt trigger

Figure 9.38 shows the package for these ICs.

Remember: any CMOS output can sink or source currents according to the power supply voltage as given by the following table:

Power Supply Voltage	Output Current (Sink or Source), Typical
5 V	0.88 mA
10 V	2.25 mA
15 V	8.8 mA

9.6   Review Questions

1. Why must a debouncing circuit be added to a switch to avoid contact problems?
2. Why are reed switches are filled with inert gas?
3. Can we use a microswitch rated at 120 V × 1 A with a load rated for 48 V × 500 mA? Why or why not?
4. How can we produce a constant-length pulse from a momentary switch signal?
5. What is the meaning of the specification "SPST" for a switch?
6. How can we use a NO switch to turn off a load?
7. What is the maximum current you can drain from a CMOS output when the IC is powered from a 10 A source?

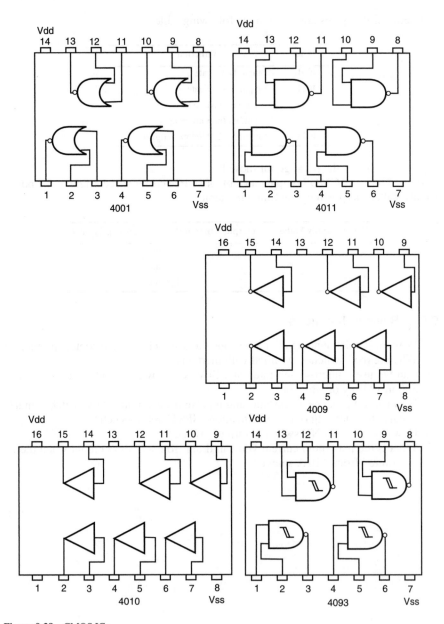

Figure 9.38 CMOS ICs.

10

Resistive Sensors

10.1 Purpose

Resistive sensors add sensory capabilities to any robotics, mechatronics, or artificial intelligence project. This chapter describes how resistive sensors operate and how they can be implemented. We will see how light-dependent resistors (LDRs) and CdS cells can be used as light sensors, adding vision to robots and other projects. In addition, negative temperature coefficient (NTC) resistors can be used as temperature sensors, potentiometers as position sensors, and conductive foam as pressure sensors, adding tactile capabilities to robots and mechanic arms. Touch sensors are also described in this chapter.

10.1.1 Theory

The world around us is filled with variable physical conditions that change continuously. Temperature, air pressure, humidity, and light are some of the conditions that fill our living environment. Any sort of robot or mechatronic device that is intended to work in this environment must have some kind of resources that allow it to interact with it. Humans have built-in sensors such as ears, eyes, etc. But how can electronic devices interact with the world?

In robotics and mechatronics designs, we use *tranducers* to interact with the external world. A transducer is a device that converts one form of energy into another. For instance, a photo cell used as a sensor can convert light level changes into electric signals. Another transducer device that can affect the ambient environment is the loudspeaker: it converts electrical energy into sound (mechanical energy).

To sense the world around them, robots and mechatronics devices can use many types of sensors. The next sections will describe some of them and describe blocks containing circuits that can be used with these sensors.

10.1.2 The LDR or CdS Cell

The *light-dependent resistor (LDR), cadmium sulfide (CdS) cell,* or *photoresistor* is a component that changes its resistance with various levels of light. In the dark, the LDR presents a very high resistance—above 1 MΩ. This resistance will fall below 100 Ω under direct sunlight. Figure 10.1 shows the appearance and the symbol used to represent an LDR.

The LDR is a resistor, so the current can flow in either direction. Although the LDR is very sensitive, "seeing" levels of light that our eyes can't, the LDR is a slow

Figure 10.1   LDR or Cds cell.

device. Fast light changes can't be detected by an LDR. The upper limit of the frequency response of an LDR is around 10 kHz. If you need to detect faster light changes, you can use other sensors such as photodiodes and the phototransistors.

Using the LDR is very easy, since it can directly bias semiconductor devices such as transistors, SCRs, ICs, etc. LDRs can be used as "electronic eyes" in applications involving robotics, mechatronics, and artificial intelligence. As in the human eye, a lens can be added to enhance the performance of an LDR in a particular application. By placing a convergent lens in front of an LDR, we can pick up more light from one direction, increasing sensitivity and adding directivity.

LDRs can be found in different sizes and formats but, in general, their electrical characteristics do not differ much. This means that almost any type can be used in the blocks described in these pages.

10.1.3   Negative Temperature Coefficient Resistors

Negative temperature coefficient (NTC) resistors, also called *thermistors* or *terminally sensitive resistors,* are components whose resistance changes with temperature. The resistance of an NTC falls as temperature increases.

Another temperature-sensitive device is the positive temperature coefficient (PTC) resistor. In this component, resistance increases with temperature. Figure 10.2 shows the symbol and appearance of common NTCs.

NTCs can be used to add temperature sensing to robots and other projects. They are specified by the resistance (in ohms) they have at a particular temperature (normally, the ambient temperature, or 20°C). Values are typically in the range from several ohms to 100 kΩ.

Temperature sensors such as the NTCs are not fast. They can't change resistance to match fast changes in temperature. This is because they have to change their own temperature, dissipating heat into the ambient, or drawing heat from it, until they reach ambient. The speed of the heat changes is basically determined by the size and the material of the NTC. Small NTCs are faster than larger ones and can better sense fast changes in ambient temperature. Because the electrical characteristics of NTCs (and PTCs) are the same as those of LDRs, they can be used with the same circuits or blocks.

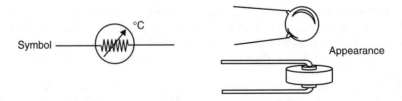

Figure 10.2   The NTC resistor, or thermistor.

10.1.4   Pressure Sensors

Simple pressure sensors can be made using conductive foam such as the materials used to protect ICs against electrostatic discharge. Figure 10.3 shows how an IC is protected by the foam and how it can be used for our purposes.

Placing the foam between two metal plates as shown in the figure, you create a pressure sensor. If you press the plates together, the resistance of the foam changes (decreases), and this can be used to trigger a circuit. You can use this kind of sensor with the same block recommended for LDRs and NTCs.

10.1.5   Potentiometers as Position Sensors

Common potentiometers can be adapted to function as position sensors. You just need to couple some kind of mechanical device (a lever or a wheel) to the shaft of a rotary potentiometer to derive an electric signal from the position of the device. Figure 10.4 shows how a rotary potentiometer can be used as a position sensor. Potentiometers with values in the range of 10 to 100 kΩ can be used as position sensors with all the blocks suggested for the other resistive sensors and described in this part.

Slide potentiometers can also be used as sensitive position sensors, as shown in Figure 10.5. See Chapter 7, Section 7.2.3, for an example of how a potentiometer can be used to sense the position of the shaft in a gearbox.

10.1.6   Touch Sensors

Two plates, in close proximity but not touching, form a touch sensor. If you place your fingers on the two plates so that you touch both at the same time, as shown in Figure 10.6, an electric current can flow across your skin, creating an electrical signal. Since the resistance of your skin is very high (100,000 Ω or more), the current flowing by this sensor isn't enough to drive common circuits. Amplification stages are needed.

An important point is that *we must keep the current low; if high voltage is used in this sensor to increase the current, electrical shock may result.*

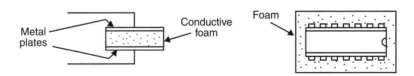

Figure 10.3   Home-made pressure sensor.

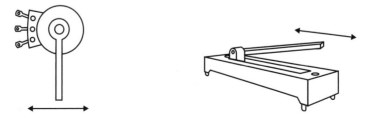

Figure 10.4   Rotary pot used as position sensor.   **Figure 10.5**   Slide pot used as position sensor.

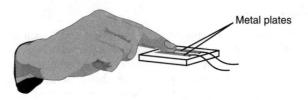

Figure 10.6   Touch sensor.

In summary, the main characteristic of touch sensors such as the ones described above is that they must operate in isolation from the ac power line and have current limit resistors. Another type of touch sensor is the one shown by Figure 10.7. If one plate is removed and connected to ground, one need only touch the other plate to trigger the circuit. The current will flow across the body of a person, activating the circuit.

10.2   How to Use Resistive Sensors

When using resistive sensors, the designer must keep in mind some important questions.

1. *How much does the resistance change under the desired operating conditions?* The resistance range of a sensor is important in the design or choice of the circuit. If a sensor changes by only few ohms resistance when excited, the circuit must have special characteristics to react to this small change. LDRs change resistance over a large range of values, so they are easy to use. NTCs, depending on the type, also have a wide resistance range, but others (e.g., pressure sensors) do not.
2. *Is the current provided by the sensor enough to drive a circuit?* If the sensor has a large resistance drop when excited, the current can be enough to drive any of our circuits. In general, to drive a block with a bipolar transistor, the resistance in the low-value condition must be less than 50,000 Ω. If Darlington transistors, CMOS ICs, or two-stage circuits are used, this resistance can be higher—in the range of 200,000 to 1,000,000 Ω. As a general rule, if your sensor doesn't trigger the circuit, try another block with higher sensitivity.

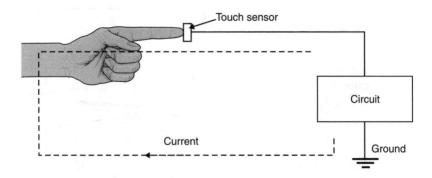

Figure 10.7   One-plate touch sensor.

10.3   Practical Blocks

The following paragraphs describe practical blocks using resistive sensors. The components of each block, in many cases, must be matched to the specific applications, depending on the sensor as well as the block it must drive.

Block 145   Basic Resistive Sensor Circuit (I)

This block can be used with LDRs to trigger a circuit when light falls on the sensor. It can also use NTCs to trigger the circuit when the temperature rises or, with a pressure sensor, to close a relay when the sensor is pressed.

The basic block is shown in Figure 10.8. In the figure, we show a simple circuit driving a 50 mA (or less) relay and the blocks that can be triggered by this circuit.

The potentiometer is used to adjust the sensitivity of the circuit. The correct value of the potentiometer must be chosen as a function of the resistance of the sensor under normal applications. The recommendation is the use of a potentiometer 3 to 5 times lower than the resistance of the sensor. Sensor with resistances below 1 kΩ are not recommended for this circuit.

Block 146   Basic Resistive Sensor Circuit (II)

The difference between this block and the previous one is that the load is triggered when the resistance of the sensor rises above a point, which is set by adjusting the potentiometer. Using an LDR, the circuit acts as a dark-activated switch. In Figure 10.9, we show the block with a simple circuit driving a relay and an indication of the blocks that can be driven by this circuit. The potentiometer must offer values between 2 and 5 times the resistance of the sensor under normal operating conditions.

Block 147   Basic Block Using a PNP Transistor (I)

The use of a PNP transistor in the driver stage, or using blocks with PNP transistors, reverses the action of the sensors, as shown by the circuit in Figure 10.10. In this example, the load is powered on when the light falling on the relay is cut. The blocks that can be used with this circuit are indicated in the figure. The potentiometer is chosen as described in Block 145.

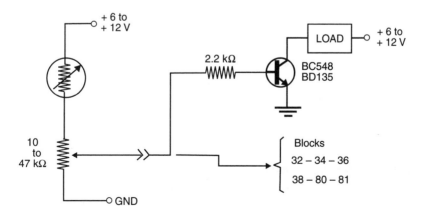

Figure 10.8   Block 145: basic resistive sensor I.

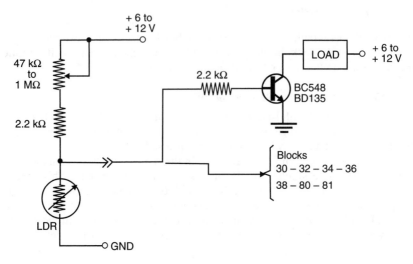

Figure 10.9   Block 146: basic resistive sensor II.

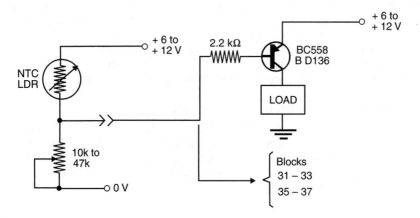

Figure 10.10   Block 147: basic block using PNP transistor I.

Block 148   Basic Block Using PNP Transistor (II)

In this block, the load is powered on when the resistance of the sensor is reduced. For an LDR, the load triggers on when the light falls on the sensor. This circuit is indicated for low-resistance sensors (below 10 kΩ). The circuit, with indications of the blocks that can be triggered, is given in Figure 10.11. The potentiometer must have a value between 3 and 10 times the resistance of the sensor under normal conditions.

Block 149   Differential Sensor

The circuit shown in Figure 10.12 operates with two sensors, comparing the resistance of each. In the LDR version, the output signal of this circuit depends on the amount of light falling on the sensors. If the amount of light simultaneously rises or falls on the two sensors, the circuit state does not change. But if the amount of light

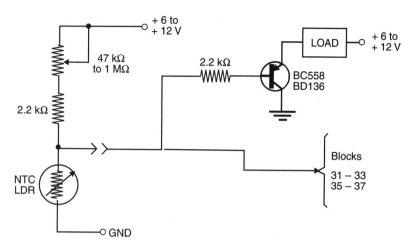

Figure 10.11   Block 148: basic block using PNP transistor II.

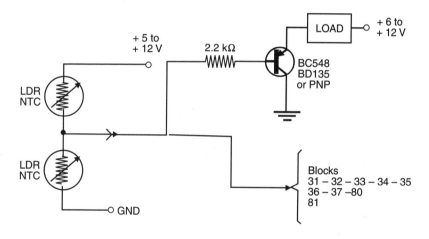

Figure 10.12   Block 149: differential sensor.

rises in LDR1 as compared to LDR2, the circuit triggers on. The potentiometer is used to adjust the trigger point of the load.

An application in robotics projects is in the "follow-the-line" robot. The same concept can be used with other sensors. Note that the circuit action is nonlinear, as one of the sensors exerts more influence on the circuit than the other. The designer must take into account the operating range of the circuit when used in a project.

Block 150   Snap Action for Resistive Sensors (I)

The slow changes in the output voltage of the previous circuit are not appropriate for drive logic circuits that use TTL or CMOS technologies. The block shown in Figure 10.13 adds a necessary snap action to resistive sensors, allowing them to better drive logic functions.

For CMOS applications, the gate can be one of the four existing in the 4093 IC. For TTL applications, the 7404 IC can be used. For CMOS applications where the sensitivity is higher (due to the high-impedance inputs), the potentiometer must be 2 to 4 times the resistance of the sensor under normal conditions, with a maximum value of 2.2 MΩ. For TTL, the same is valid, but the upper resistance limit is about 47 kΩ due the sensitivity, which is lower than that of the CMOS gates.

Block 151   Snap Action for Resistive Sensors (II)

The output of the block shown in Figure 10.14 is low when the sensor, if an LDR, is in the dark. When the sensor is illuminated, the output of the circuit passes to the high logic level. The circuit has the snap action needed to drive logic blocks, since a Schmitt inverter is used.

For CMOS applications, a logic inverter can be made using one of the four gates of a 4093, and for TTL applications the 7402 can be used. The same characteristics described in the previous blocks are found in this block.

Block 152   Increasing Sensitivity

The simple block shown in Figure 10.15 is used to increase the sensitivity of a resistive sensor if it isn't able to generate enough current to drive blocks such as the ones

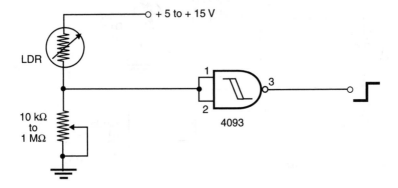

Figure 10.13   Block 150: snap action CMOS.

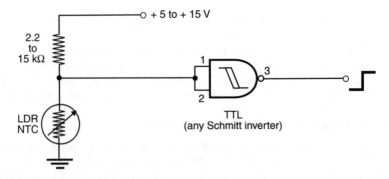

Figure 10.14   Block 151: snap action TTL.

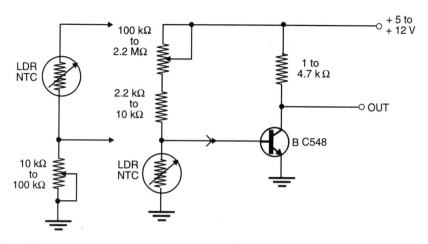

Figure 10.15   Block 152: increasing sensitivity.

shown in this book. The use of a block like this one is necessary if, in the low-resistance condition of the sensor, the resistance still is not low enough to provide the current needed by the circuit. For instance, if you are using this block with an LDR and want to detect very weak light sources, you must amplify the signal before this block can be used. In general, this block is recommended when the resistance of a sensor in an application varies in the range from 100 kΩ to infinity.

Resistor R is chosen to give the desired circuit sensitivity. The ideal values are between 2 and 5 times the value of the sensor in the low-resistance condition. In the diagram, we show the blocks in which this circuit can be used to drive the circuit.

Block 153   Light-Activated Circuit Using an SCR

Sensitive SCRs such as those of the 106 series (TIC106, IR106, MCR106, C106, etc.) can be triggered by short light pulses to drive loads such as solenoids, relays, motors, etc. The block shown in Figure 10.16 shows how this can be done.

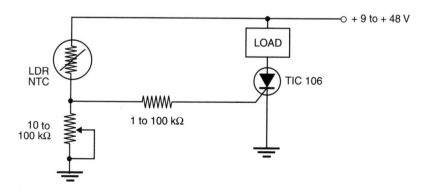

Figure 10.16   Block 153: SCR block.

The potentiometer needs to have a value about 2 to 5 times lower than the resistance of the sensor under normal (not excited) conditions. This potentiometer is used to adjust the triggering point of the circuit. Three important points must be taken into account in this configuration:

1. The SCR is a latching device when powered from dc sources. Once the SCR is on, it remains on even after the trigger pulse disappears.
2. A voltage drop of about 2 V across the SCR appears when it is in the on state. You must compensate for this voltage drop by increasing the power supply voltage.
3. To trigger the circuit off, it is necessary to turn off the power supply or add a momentary contact switch between the anode and the cathode of the SCR.

The same concepts used when an LDR is installed are valid for other resistive sensors. If NTCs are used, this circuit can function as an overtemperature switch.

Block 154   Dark-Activated Circuit Using an SCR

The block shown in Figure 10.17 turns on a load (relay, motor, etc.) when the amount of light falling onto the LDR is reduced by a shadow or if the sensor moves away from a light source.

The potentiometer in this circuit must have between 2 and 5 times the resistance of the sensor under normal operating conditions (without light). This potentiometer adjusts the trigger point of the circuit. The points discussed in the previous block about circuit operation are also valid here.

Block 155   Priority Circuit Using Resistive Sensors

This circuit can be used in many practical projects involving robotics, mechatronics, and artificial intelligence. If the sensors are LDRs, the circuit shown in Figure 10.18 triggers one of the loads, depending on which LDR was illuminated first.

Applying two flashes of light, with one falling on each LDR, the first flash triggers on the corresponding SCR, causing a voltage drop in the circuit that biases the other sensor. When using the other sensor, it is important remember the activation speed of the circuit. As in the case of a temperature sensor, the action is very slow, as it depends on the temperature rise of each sensor.

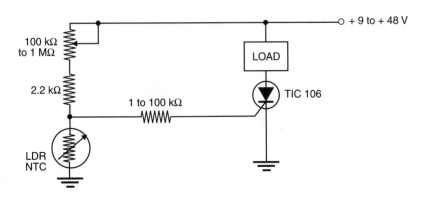

Figure 10.17   Block 154: dark-activated SCR.

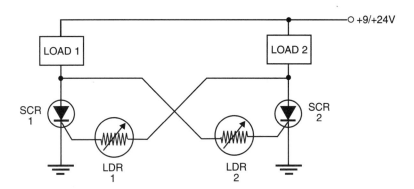

Figure 10.18   Block 155: priority circuit using LDRs.

Block 156   Opto-Isolator Using LDR

Figure 10.19 shows how a simple opto-isolator can be made using a small incandescent lamp and an LDR. This configuration can be used to isolate one block from another in a robotics or mechatronic project. For instance, the lamp can be driven by a low-voltage circuit, and the LDR can be used to control a high-voltage circuit.

Both LDR and the lamp, must be installed inside an opaque enclosure to avoid the influence of the ambient light.

Any small lamp with currents in the range between 10 and 500 mA can be used. You can also use neon lamps in series with a 100 kΩ resistor to drive the circuit from the ac power line. In the 220/240 Vac power line, place a 220 kΩ resistor in series with the lamp. In the figure, we indicate blocks that can be driven by this circuit.

Block 157   Opto-Isolator with Logic Input (TTL and CMOS)

Figure 10.20 shows how an LED can be used to drive an opto-isolator from TTL or CMOS outputs. The value of resistor R depends on the logic and the power supply voltage as given by Table 10.1.

The LED and the LDR must be installed inside an opaque enclosure to avoid the influence of ambient light. Remember that the resistance of the LDR falls when the input logic level is high (1).

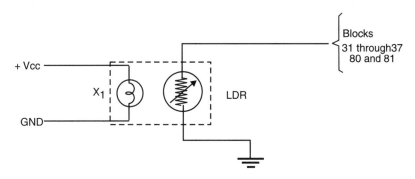

Figure 10.19   Block 156: opto-isolator using LDR.

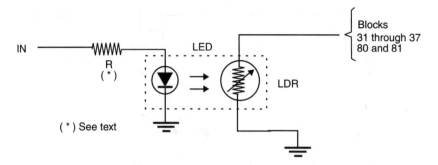

Figure 10.20   Block 157: opto-isolator for TTL logic.

Table 10.1   R Value vs. Logic and Voltage

Logic	Value of R
TTL	330 Ω
CMOS-5 V	390 Ω
CMOS-6 V	470 Ω
CMOS-9 V	820 Ω
CMOS-12 V	1 kΩ
CMOS-15 V	1.5 kΩ

Note: Complete devices using LEDs as emitters, and devices such as Schmitt triggers, SCRs, transistors, Darlington transistors as receivers, and other configurations can be designed.

Block 158   Current Sensor Using an NTC

The block shown in Figure 10.21 can be used to sense the current across a circuit without making electric contact. The nicrome wire is placed near an NTC inside a small enclosure. When the temperature of the wire rises due the amount of current across it, the NTC senses the changes and can act on the circuit. This circuit can be used to trigger a protective circuit when a dc motor is stalled by an obstacle it has encountered, thereby avoiding overload problems.

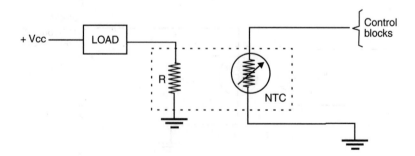

Figure 10.21   Block 158: current sensor.

Block 159   Thermal Crowbar

This circuit can be used to protect motors and other loads in case of overload. The NTC is mounted near a nicrome wire or a wire-wound resistor. A fuse is used as a protective element as shown in Figure 10.22.

When the temperature of the resistor or wire rises because of an overcurrent condition, the NTC triggers the SCR on, and the fuse is put in a short circuit with ground. This causes the fuse to open, protecting the circuit.

The resistance of the potentiometer must be 3 to 5 times the resistance of the NTC in the normal operating temperature. NTCs with resistances in the range between 1 and 100 kΩ are recommended for SCRs of the 106 family. See the crowbar circuits described in previous blocks for more details. Remember that the SCR can handle the high current necessary to burn the fuse in this application.

Block 160   Light/Temperature-Controlled Oscillator

The circuit of Figure 10.23 is an analog-to-digital converter (ADC or A/D converter); the amount of light falling onto a sensor, or the temperature of a sensor, is converted into an oscillation frequency. The sensor's resistance and capacitor C1 determine the output frequency range. With the values shown in the figure, the circuit produces an audible tone that is emitted by a loudspeaker.

Applications for this circuit in robotics and mechatronics include an audible reaction from a robot when it touches hot objects or when it "sees" a flash of light. By adding potentiometer P1, is possible to put the oscillator on the threshold of oscillation so that it is triggered by light falling onto the sensor or when the temperature rises above a certain value.

P1 and the sensor can be interchanged to invert the action of the circuit (i.e., it is triggered by darkness or when the temperature falls). This circuit also operates with other resistive devices such as touch sensors and pressure sensors.

Block 161   Light-Sensitive/Temperature-Dependent Oscillator

To increase the sensitivity of an oscillator using a CMOS NAND gate, you can add a transistor as shown in Figure 10.24. The capacitor determines the frequency range, and resistor R1 determines the center frequency. This is a variable-duty-cycle oscillator in which the sensor acts only within the width or separation of the pulses, according to whether you use a PNP transistor or an NPN device and invert the diode.

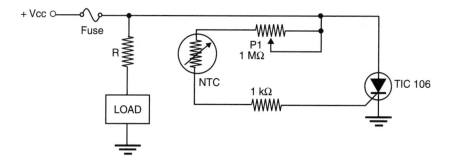

Figure 10.22   Block 159: thermal crowbar.

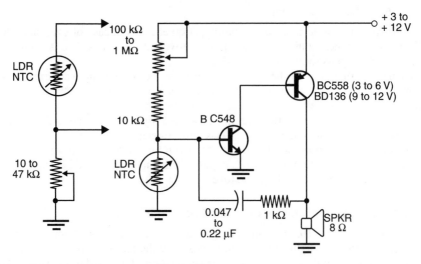

Figure 10.23   Block 160: sensor-controlled oscillator.

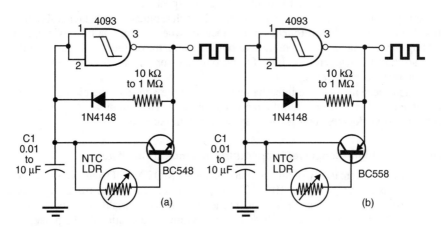

Figure 10.24   Block 161: light/temperature controlled oscillator.

Block 162   Light/Temperature-Dependent Oscillator

The block shown by Figure 10.25 is another configuration of an oscillator in which the frequency depends on the temperature or the amount of light on a sensor. The circuit uses two CMOS NAND (or NOR) gates. In this configuration, we show the 4001 IC, but the 4011 can also be used without problems. The capacitor will determine the frequency range. The circuit can drive many of the blocks described in this book and can be powered from 5 to 15 V power supplies.

Block 163   Light/Temperature-Triggered Monostable

The circuit shown in Figure 10.26 can be triggered when the resistance of a sensor falls (a) or rises (b). Once triggered by a short burst of light (or darkness), the output

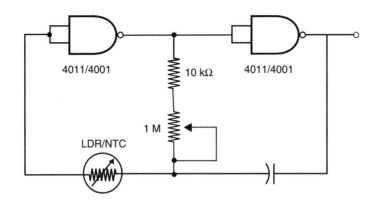

Figure 10.25   Block 162: temperature/light-dependent oscillator.

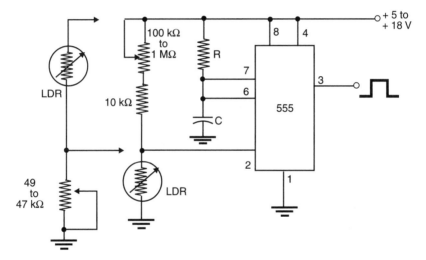

Figure 10.26   Block 163: light/temperature-triggered monostable.

of the 555 IC goes to the high logic level and can deliver voltage to a load. The output remains on for a time interval determined by R and C. The resistor can be a 10 kΩ or higher unit, and C must have values in the range between 1 nF and 1,000 μF. Using a 1,000 μF capacitor and a 1 MΩ resistor, the circuit, once triggered, will remain on for about 15 min.

The circuit can be powered from supplies in the range between 5 and 18 V. If we use the TLC7555 (the CMOS version of the 555), the circuit can be powered starting from 3 V. There is also a very low voltage version of the 555 that can operate with voltage supplies as low as 1.5 V.

Block 164   Fast Monostable Sensor

This circuit is recommended to detect fast pulses of light if the sensor is an LDR. The operation is the same as in the previous block. The components shown in Figure 10.27 are also determined as in the previous block.

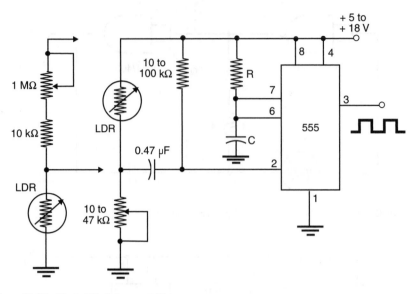

Figure 10.27   Block 164: fast monostable.

Block 165   Light/Temperature-Dependent Oscillator

This circuit can also be used with touch or pressure sensors. The block shown in Figure 10.28 produces a squarewave signal in the range from a fraction of a hertz to 500 kHz, depending on R and C in the circuit. Frequency is calculated using the formula in the diagram. By replacing one of the branches of potentiometer P1 in the circuit shown in Block 92, it is possible to have a temperature/light-controlled pulse width modulation (PWM) circuit.

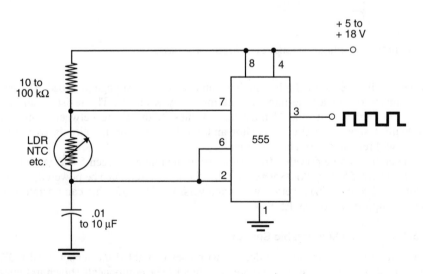

Figure 10.28   Block 165: light/temperature-dependent oscillator.

10.4   Additional Information

Figure 10.29 shows the spectrum of some light sources and the LDR. These curves can be useful when using an LDR such as an IR or UV sensor. Other nonresistive sensors such as photodiodes, photo cells, and phototransistors are included.

10.5   Suggested Projects

The potentiometer in many blocks recommended for the control of dc motors can be replaced by resistive sensors. In particular, we suggest:

1. Blocks 65 and 66: replace the potentiometer with an LDR or NTC to create a light- or temperature-controlled load.
2. Block 70 through 74: replace the potentiometer with resistive sensors.
3. Create a configuration that makes a robot follow a dark stripe on the ground.
4. Design a robot that uses two LDRs to locate a light source such as a lamp or a candle.
5. Design a system as above, and add intelligence so that the robot that looks for a lamp, but make it "afraid" of excessive light so that it stays a "safe" distance away from the source. Figure 10.39 shows a circuit that can be upgraded for this task.
6. Create an arm that can be controlled by potentiometers, using other potentiometers as position sensors.
7. Add a temperature sensor to a robot that can separate hot objects from cold ones.

10.6   Review Questions

1. What is a transducer?
2. When exposed to light, what happens to the resistance of a CdS cell?
3. Temperature sensing can be added using what kind of sensor?
4. Why are small NTCs are faster than large ones when sensing temperature changes?
5. How can a robot see in the dark using an LDR?

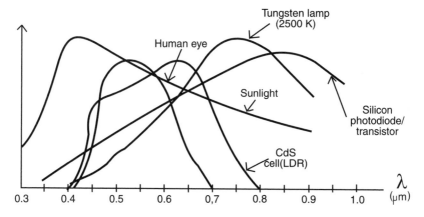

Figure 10.29   Spectral performance for some sources/sensors.

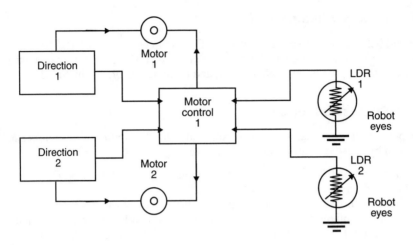

Figure 10.30   Robot that is "afraid" of light.

11

Operational Amplifiers and Comparators

11.1 Purpose

The purpose of this chapter is show the reader how operational amplifiers and comparators can be used in projects involving robotics and mechatronics. This chapter shows how a basic operational amplifier functions and how it is used as voltage comparator. Following the explanatory material, many blocks using operational amplifiers and comparators will be described, driving loads such as relays, motors, and solenoids, and operating with signals received from sensors such as LDRs, NTCs, and touch sensors.

11.2 Theory

Operational amplifiers have a wide range of applications in today's electronics. They can boost signals from sensors, operate as filters, and generate signals. The robotics and mechatronics designer can find many commercial types that are suitable for these projects. However, to choose a suitable device for a given application, you must know how it functions and be able to interpret its main characteristics.

11.2.1 Operational Amplifiers and Comparators

Operational amplifiers (opamps) are circuits that were originally designed to perform mathematical operations in analog computers. But because they are sensitive amplifiers that can operate in the range of dc to medium frequency signals, they soon found many other applications.

Today, the designer of robotics and mechatronics projects can find many opamps in the form of integrated circuits (ICs). These ICs are cheap, easy to find and use, and possessed of some common electrical characteristics that make them ideal for projects in the areas addressed by this book.

The basic opamp has two inputs: the inverting input (–) and the noninverting input (+). If a signal is applied to the inverting input, it appears with the opposite phase in the output, as shown by Figure 11.1. On the other hand, if it is applied to the noninverting input, it appears with the same phase at the output.

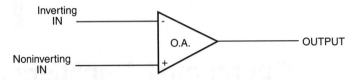

Figure 11.1   An operational amplifier.

The basic electrical characteristics of an operational amplifier are as follows:

- Very high input impedance (several megohms to more than 1,000,000 MΩ in types using FETs)
- Very low output impedance (50 to 100 Ω, typically)
- High voltage gain (10,000 to more than 1,000,000)

The gain of an operational amplifier in an application is given by the feedback loop as shown by the Figure 11.2.

An important application for the opamp is the *voltage comparator.* A voltage comparator is a high-gain operational amplifier that compares two input voltages and delivers an output that indicates when the inputs are equal or unequal.

In one of the opamp inputs, we apply a reference voltage that is normally provided by a voltage divider that is formed by resistors as shown in Figure 11.3. The output is one of the following three possibilities:

- If the voltage applied to the input is equal to the reference voltage, then the output is zero.
- If the voltage applied is higher than the reference, the output is a positive voltage.
- If the input voltage is lower than the reference, the output is a negative voltage.

The comparator, because of its high gain, has a "snap" action when the input passes across the reference voltage, as shown by the curve in Figure 11.4. This snap action makes the comparator ideal for applications involving sensors, as we will explore in this chapter.

It is also possible to apply the reference voltage to the noninverting input. In this cases, the output is inverted in comparison to the input signal; that is, the output goes low when the voltage rises above the reference.

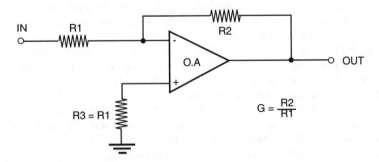

$$G = \frac{R2}{RT}$$

Figure 11.2   Gain of an opamp.

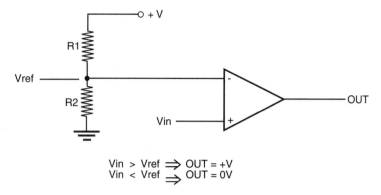

$$Vin > Vref \Rightarrow OUT = +V$$
$$Vin < Vref \Rightarrow OUT = 0V$$

Figure 11.3   The voltage comparator.

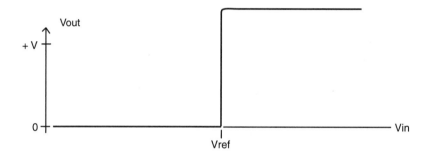

Figure 11.4   Snap action of a voltage comparator.

11.2.2   The Window Comparator

Two comparators can be connected as shown in Figure 11.5 to form the so-called *window comparator.* Its operation is very simple to explain: one of the comparators triggers with the reference voltage V1, and the other with the reference voltage V2, determined by the voltage divider formed by R1, R2, and R3. In this case, V2 is higher than V1.

Assuming that the input voltage is zero, comparator 1 (VC1) has its output in the high logic level or the power supply voltage, and the output of comparator 2 (VC2) is at 0 V. Considering the presence of the diodes in the output, combining the voltages of OUT1 and OUT2, the voltage level in OUT is high.

When the input voltage rises to V1, VC1 changes its output to 0 V. As a result, both outputs (OUT1 and OUT2) are now zero, and the combined output is zero.

The voltage continues to rise, when it reaches V2, comparator 2 (VC2) changes its output from 0 V to a positive value near the power supply voltage. Combining the two outputs, we have now an OUT positive. Figure 11.5a shows this concept in graphic form. The presence of the diodes is important, since they block the circulation of current from one output to the other when one is positive and the other is negative.

In the first segment (0 to V1), the output is positive. In the second segment (V1 to V2), we have a "window" in which the voltage falls to zero. Continuing, in the third

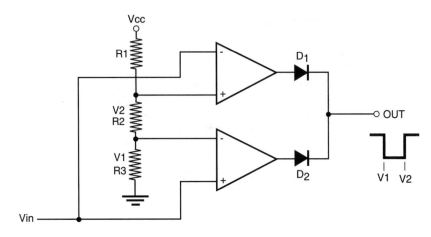

Figure 11.5   The window comparator.

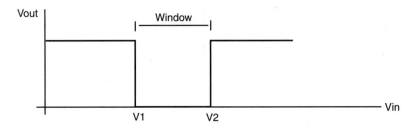

Figure 11.5a   Characteristics of the window comparator.

segment (V2 and more), the output is again positive. Note that this circuit can be used to detect voltages in a range determined by V1 and V2. The values of R1, R2, and R3 will determine the width of the window.

If the input comes from a sensor, the circuit can be used in "intelligent" applications, as sensors can recognize light and temperature levels, and other conditions, in a predetermined range. As we also see in the following pages, it is possible to change the window, adding more intelligence to the circuit.

Another operational possibility for the window comparator is shown in Figure 11.6. In this case, we have a positive window instead of the negative windows described previously. The output of this circuit is high or positive in the interval between V1 and V2.

11.2.3   How to Use Operational Amplifiers and Comparators

The operational amplifiers can be used in two basic forms:

1. *Linear mode.* In the linear mode, the opamp operates as a common ac or dc amplifier with a voltage gain given by the feedback loop. This operating mode is chosen when signals must be amplified without changes in the waveform, as shown by Figure 11.7.

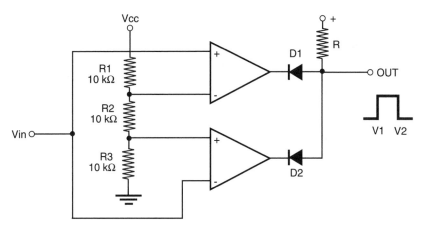

Figure 11.6   Another window comparator.

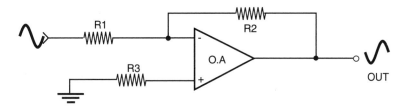

Figure 11.7   Operation in the linear mode.

2. *Comparator.* When used as a comparator, the opamp has a snap action, moving very quickly from the high to the low or, low to the high, logic level in the output when the input changes slowly. This operating mode is preferred when the circuit is used to trigger a load from signals sent from sensors or other circuits.

11.2.4   Choosing an Opamp

When looking for an operational amplifier for a particular application, the designer must consider the following characteristics.

Power supply voltage.   Opamps can be found in voltage ranges from 1.5 to 30 V. Some types don't operate well with low-voltage supplies; e.g., the 741 (the most popular) needs at least 9 V. The designer must be certain that the voltage in the circuit is high enough to power the opamp.

Open-loop gain.   This is the maximum gain of an opamp, which varies from 10,000 to 100,000. Only when used as comparators can opamps operate with the maximum gain. When used in the linear mode, the gain is determined by the feedback loop. In the component's data sheet, the gain can be given in dB or in V/mV or μV/V.

Voltage gain × bandwidth.   While the voltage gain rises, the maximum frequency amplified by an opamp is reduced. The frequency at which the voltage gain falls to 1 determines the bandwidth of an opamp, or the maximum frequency that can be amplified by the operational amplifier. Common types such as the 741 have a low bandwidth of approximately 1 MHz. This fact makes these components unsuitable for high-frequency applications. The maximum operational frequency of an opamp is also referred to as f_T (transition frequency).

Common mode rejection ratio (CMRR).   When the same voltage is applied to the inverting and noninverting inputs, the output must be zero. The ability to reject signals of the same amplitude applied to the inputs determines the CMRR. The CMRR is specified in dB, and typical values are around 90 dB.

Input impedance.   Some types of operational amplifiers have very high input impedance. For the bipolar types, the input impedance is in the range of a few megohms. In types using FETs, this impedance can rise to several billions of megohms or higher.

Output impedance.   The output impedance of an operational amplifier is very low—normally in the range from 50 to 250 Ω for popular types. This impedance, specified in ohms, indicates also the amount of current that can be delivered or drained by the operational amplifier.

11.2.5 Power Supplies

Opamps can be powered from single- or dual-voltage sources as shown in Figure 11.8. When the devices are powered from single voltage supplies, the output voltage swings from 0 V to the power supply voltage. When a dual supply is used, the output voltage swings from −V to +V. A dual power supply suitable for applications in this book is given in Figure 11.9.

The transformer has a primary according to the ac power line. The secondary depends on the output voltage. Table 11.1 gives the voltages of the transformer, the recommended ICs, and the output voltage.

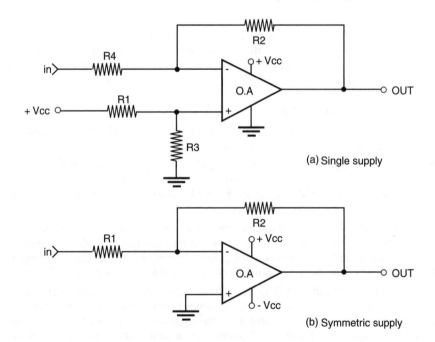

Figure 11.8   Powering an opamp.

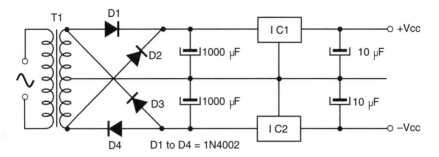

Figure 11.9   Dual-voltage power supply.

Table 11.1   **Transformer Voltage, Recommended ICs, and Output Voltage**

Transformer*	IC1	IC2	Output Voltage
6 V × 1 A	7805	7905	5 + 5 V
7.5 V × 1 A	7806	7906	6 + 6 V
9 V × 1 A	7809	7909	9 + 9 V
12 V × 1 A	7812	7912	12 + 12 V
15 V × 1 A	7815	7915	15 + 15 V

*The transformer current can assume any value from 100 mA to 1 A. The current assumed by the transformer will be the current delivered to the output.

11.3   Practical Blocks

The following are practical blocks using resistive sensors, opamps, comparators, and window comparators. In many cases, block components must be matched to the specific applications, depending on the sensor used and also on the block it must drive.

The blocks are designed to use common operational amplifiers such as the 741, LM324, CA3140, and many others. Depending on the type of opamp used, variations in the other components may be necessary to achieve desired performance.

Block 166   Voltage Follower

In the voltage follower, the voltage gain is 1. The voltage of the output signal is the same as the signal applied to the input. The advantage in using the configuration shown in Figure 11.10 is that the circuit converts a very high input impedance into a very low output impedance, resulting in a real power gain to the block.

The opamp for this configuration can be a 741, LM324, or any other, depending on the applications and the power supply voltage. If the circuit is intended to operate with negative voltages, a symmetric or dual power supply is needed.

Block 167   Amplifier with Gain

The voltage gain of an operational amplifier is given by the feedback network formed by resistor R2 in the circuit shown in Figure 11.11. The resistance ratio be-

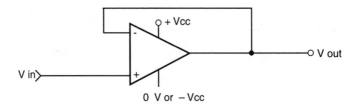

Figure 11.10   Block 166: voltage follower.

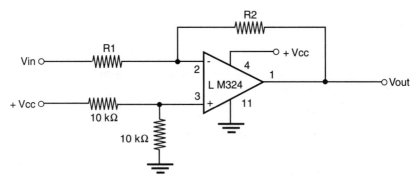

Figure 11.11   Block 167: amplifier with gain.

tween this resistor (R2) and R1 gives the voltage gain. In the circuit shown, we can program the gain in a wide range of values according to the following table:

R2	Gain
10 kΩ	2
100 kΩ	10
500 kΩ	50
1 MΩ	100
10 MΩ	1000

The circuit can be powered from single supply, but the voltage divider, applying half of the supply voltage to the noninverting input, is necessary. The resistor can range from 10 to 100 kΩ in this input, depending on the application.

Block 168   Driving an NPN transistor

The block shown in Figure 11.12 shows how a comparator can be used to drive an NPN transistor that controls a relay. This block can be used to drive other loads as suggested in the figure. The figure lists blocks that can be driven by the comparator.

In the application, the relay is on (or the load is powered) when the output of the comparator goes to the HI logic level. The relay or load can be powered from a voltage source that differs from the one used for the comparator. It is only necessary to have a common ground. Common uses for this circuit are to drive loads from low-power signals such as those coming from sensors.

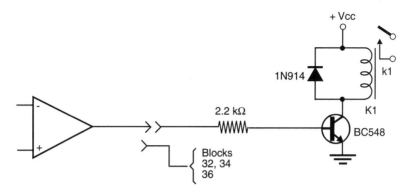

Figure 11.12   Block 168: driving an NPN transistor.

Block 169   Driving a PNP Transistor

The block shown in Figure 11.13 drives a PNP transistor from common operational amplifiers or a comparator. This block can be used to drive other blocks such as the ones indicated in the figure. As in the previous block, the transistor and comparator can be powered from different voltage sources. It is only necessary to use a common ground.

Block 170   Basic Voltage Comparator

Figure 11.14 shows the basic voltage comparator. This circuit can be used with sensors or other signal sources to drive relays or other circuits. See in Block 167 or Block 168 how to drive loads (relays and other loads) from this block.

In a robotics or mechatronics application, this circuit can be used with resistive sensors or controls. The trigger point, or the voltage in which the output of the comparator goes from the low logic level to the high logic level (the power supply voltage), is adjusted by the voltage divider formed by R1 and R2. For instance, if R2 is adjusted for 10 kΩ, the trigger point will be half of the power supply voltage. See in Section 11.2, Theory, how to calculate the trigger point for this circuit. Remember that the input voltage never can be higher than the power supply voltage, and the circuit has a high input impedance.

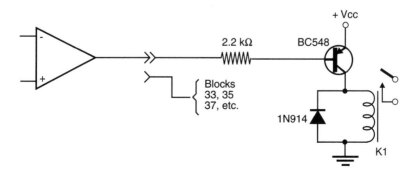

Figure 11.13   Block 169: driving a PNP transistor.

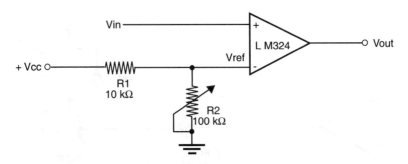

Figure 11.14   Block 170: the basic voltage comparator.

The values of R1 and R2 can assume a wide range of resistances. Values between 1 kΩ and 1 MΩ can be used experimentally according to the input characteristics of the installed comparators.

Block 171   Negative Voltage Comparator

A negative voltage comparator is shown in Figure 11.15. The output of this circuit goes from the high logic level to the low logic level when the input voltage rises to a value higher than the reference voltage.

The trigger point is determined as explained in Section 11.2, Theory, depending on the adjustment of R2. The values for R1 and R2 can be chosen in the same range as suggested in the previous block. This block can drive loads such as relays, motors, and so on, as described in Blocks 167 and 168.

Block 172   Using Resistive Sensors (I)

A complete block for a resistive sensor (LDR or NTC) driving a relay is shown in Figure 11.16. In configuration (a), when the amount of light at the sensor decreases, the voltage at the input of the comparator rises. When the reference voltage (fixed by R1 and R2) is reached, the comparator triggers, with the output voltage passing from zero to the positive value of the power supply. This causes the transistor to go to a saturated state, driving the relay. The same circuit can be used for temperature sensing if an NTC is used as the sensor. In configuration (b), the inverse occurs, with the transistor turning on the relay (or other load) when the amount of light on the LDR increases. The trigger point can be adjusted by P1. This circuit can drive other blocks as suggested in Blocks 167 and 168.

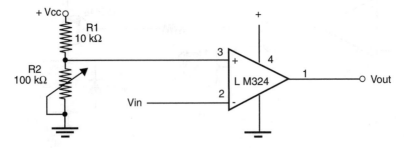

Figure 11.15   Block 171: negative voltage comparator.

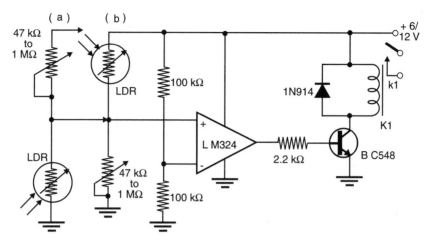

Figure 11.16   Block 172: using resistive sensors I.

Block 173   Using Resistive Sensors (II)

Figure 11.17 shows another way to use a cooperator with resistive sensors as LDRs and NTCs. In configuration (a), the relay triggers when the amount of light increases beyond the adjusted level, and in (b), when the amount of light decreases in this manner. The same is valid for a resistive sensor. This block can also be used to drive other loads as suggested in Blocks 167 and 168.

Block 174   Resistive Sensors Driving PNP Transistor

Figure 11.18 shows how a PNP transistor can be used to drive a relay with a comparator where the sensor is resistive. The difference is that, in this configuration, the relay is activated when the output of the comparator passes from the high logic level to

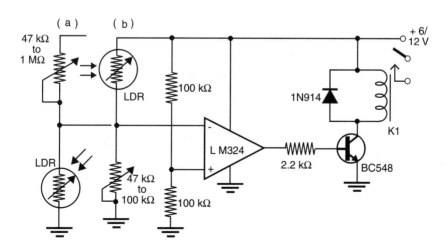

Figure 11.17   Block 173: using resistive sensors II.

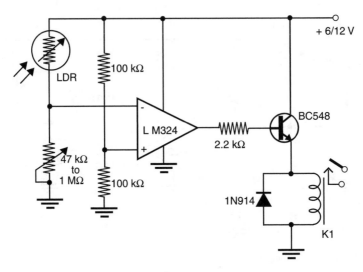

Figure 11.18 Block 174: driving PNP—negative.

the low logic level. The same circuits suggested in Blocks 167 and 168 can be driven from this block.

Block 175 Differential Sensor

The block shown in Figure 11.19 offers interesting application options to the robotics and mechatronics designer. The relay triggers when the resistance of one of the sensors rises or when the resistance of the other sensor falls. If the resistance of both sensors falls or rises at the same time, the circuit doesn't trigger.

The circuit can be used as a differential light sensor, temperature sensor, or moisture sensor. P1 adjusts the trigger point of the circuit, and the values of the resistors

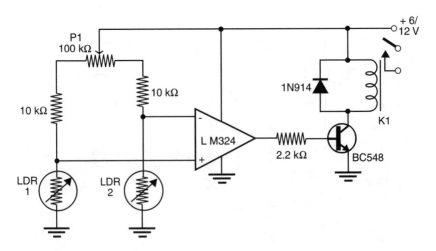

Figure 11.19 Block 175: differential sensor.

depend on the resistance of the sensors. It is recommended that these resistors have resistances of the same value as the sensors in normal operating conditions.

Block 176   Touch Sensor/Pressure Sensor

A highly sensitive touch sensor is shown in Figure 11.20. The circuit triggers the relay when the sensor is touched. You can replace the touch sensor with a pressure sensor or other resistive transducer. Resistor R1 must have a value that is appropriate for the desired sensitivity. For touch applications, the value is typically between 1 MΩ and 10 MΩ (higher values give higher sensitivity). If the positions of the sensor and R1 are reversed, the circuit can be used as a turn-off sensor. Blocks 167 and 168 suggest loads that can be controlled by this circuit.

Block 177   Delayed Turn-On Relay

In the block shown in Figure 11.21, the relay turns the circuit on after a delay. The delay time can be adjusted by P1 and depends on the capacitor. Using a 1,000 μF ca-

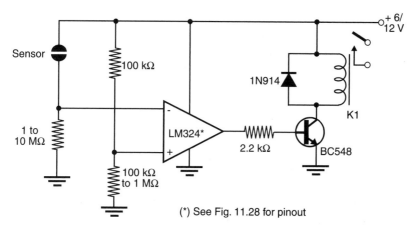

Figure 11.20   Block 176: touch sensor.

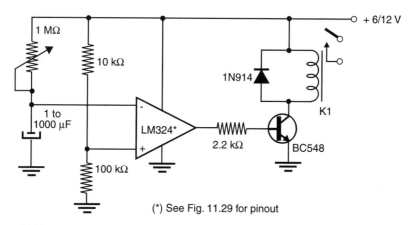

Figure 11.21   Block 177: delayed relay.

pacitor and a 1 MΩ variable resistor, the time delay can extend to intervals up to 30 min. Other loads can be controlled by this block as shown by Blocks 167 and 168.

Block 178   Driving TTL Blocks

The block shown in Figure 11.22 shows the correct use of an LM324 when driving TTL loads. The operational amplifier (as comparator) can be powered from a 5 V supply.

Block 179   Power Comparator

Figure 11.23 shows how an operational amplifier can directly drive a high-power load using a medium-power transistor such as the BD135. This circuit can drive loads up to 500 mA with a voltage gain of 10. This means that changes in the input voltage between 0 and 0.6 V will cause changes in the output voltage in the range between 0 and 6 V (the circuit must be powered with voltages higher than the maximum voltage desired in the output). The output transistor must be mounted on a heat sink.

Block 180   Low-Frequency Squarewave Oscillator

The block shown in Figure 11.24 can produce low-frequency signals in a range that depends on the opamp (typically between 0.001 Hz and 10 kHz). The frequency is determined by R1 and C1. Applications in robotics and mechatronics include tone generators for remote controls, signaling, or to flash a lamp (with an added high-power output stage). Using an LM324, the circuit can produce signals of up to 1 MHz.

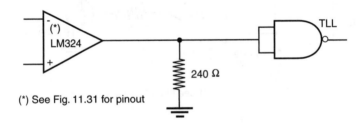

Figure 11.22   Block 178: driving TTL gates.

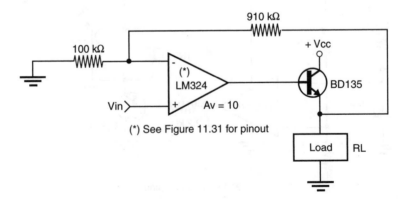

Figure 11.23   Block 179: power comparator.

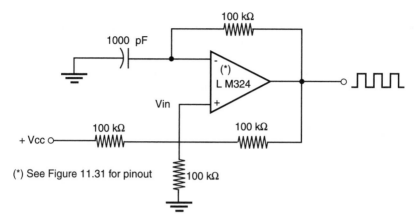

Figure 11.24   Block 180: low-frequency squarewave oscillator.

Block 181   Double Comparator

Two comparators can be connected to drive two loads with different levels of voltages as shown in Figure 11.25. The voltage divider formed by R1, R2, and R3 determines the reference voltages V1 and V2. When the input voltage reaches V1, the first comparator turns on, driving the output. When the input voltage reaches V2, it is the second comparator that turns on its load. Any of the blocks suggested in Blocks 167 and 168 can be wired to this circuit. One or more resistor of the voltage divider can be adjusted, depending on the application.

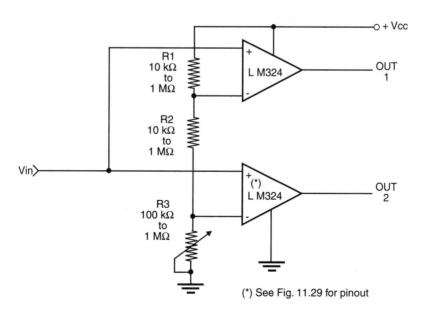

Figure 11.25   Block 181: double comparator.

An interesting application for this circuit is in intelligent robots. The input signal can be delivered by an LDR (see the circuits in Chapter 10). The first comparator controls a motor, and the second controls a reversing system. When the light picked up by the sensor (LDR) is weak, the robot advances until it becomes strong enough to trigger the second comparator. The relay plugged into this comparator reverses the motor for a few seconds and stalls the robot, keeping it away from the light source.

Block 182   Step Comparator

Many comparators are designed to trigger with progressive voltages. Such a device is a step comparator as shown in Figure 11.26. This circuit uses four comparators, but an LM324 can be used to add intelligence or complex functions to a robot or a mechatronic device.

The outputs of the comparators go to the high logic level when the voltage rises to a level between zero and the maximum (i.e., positive power supply voltage). Any of the blocks suggested in Blocks 167 and 168 can be driven by this circuit. The resistors in the voltage divider can have values from 1 to 100 kΩ, and it is recommended that you keep the sum of them below 1 MΩ.

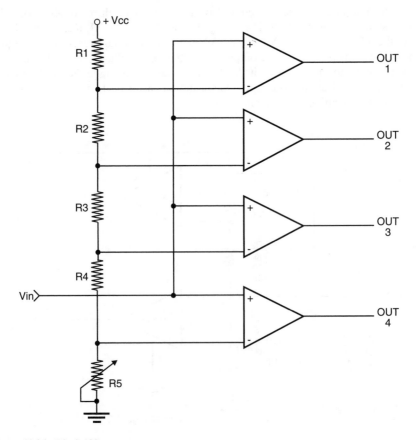

Figure 11.26   Block 182: step comparator.

Block 183   Window Comparator (I)

Figure 11.27 shows a simple window comparator using two of the four comparators that are available in an LM324. The output of this circuit remains at 0 V when the input is in the range between V1 and V2. V1 and V2 can be calculated by the following formulas:

$$V1 = R\frac{1 \times Vcc}{R1 + R2 + R3}$$

$$V2 = \frac{(R1 + R2) \times Vcc}{R1 + R2 + R3}$$

where   V1 and V2 = reference voltages in volts
        R1, R2, and R3 = resistances in ohms
                Vcc = power supply voltage

R1 to R3 can assume values in the range between 1 kΩ and 1 MΩ. It is recommendable that R1 + R2 + R3 < 1 MΩ. The output of this block can be plugged into any of the loads suggested in Blocks 167 and 168.

Block 184   Window Comparator (II)

The output of the window comparator shown in Figure 11.28 remains high when the input voltage is kept between V1 and V2. The operating principle is as described in Section 11.2, Theory. The values of V1and V2, which determine the window, are calculated as described in the previous block. The same blocks that can be driven by Blocks 167 and 168 can be driven by this circuit.

The ideal IC for this application is the LM324, but other operational amplifiers or comparators can be used experimentally. Observe the proper characteristics of the type used. Resistive sensors or other circuits can deliver signals to this circuit. Remember that the input voltage must be kept below the power supply voltage.

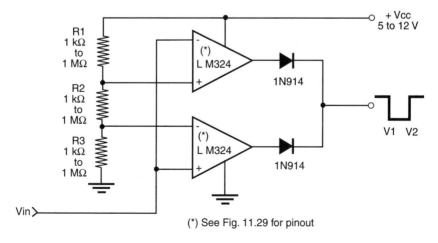

Figure 11.27   Block 183: window comparator I.

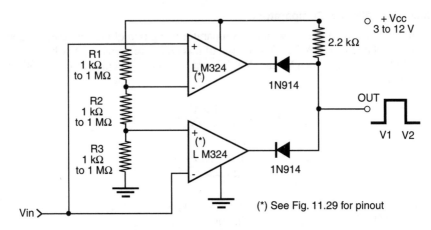

Figure 11.28   Block 194: window comparator II.

11.4   Additional Information

The following describes some operational amplifiers and comparators that are suitable for the applications described in this chapter.

741.   This is the most popular of all operational amplifiers and is suitable for almost all of the blocks described in this book. The 741 can appear with designations or suffix according to the manufacturers as: LM741, uA741, LH101, MIC741, SN72741, MC1439, etc. Versions with two or four amplifiers equivalent to the 741 in the same package are available.

Characteristics
- Power supply voltage: 9 to 36 V (single)
- Open loop gain: 100 dB (30 µV/V)
- Input impedance: 1 MΩ
- Output impedance: 150 Ω
- f_T (transition frequency): 1 MHz
- CMRR: 90 dB

LM324.   This IC contains four operational amplifiers that can be used as comparators in many applications.

Characteristics
- Power supply range: 3 to 25 V (single)
- Open loop gain: 100 V/mV (100 dB)
- f_T (transition frequency): 1 MHz

TL070.   This is a JFET operational amplifier (versions with two and four amplifiers in the same package are available).

Characteristics
- Power supply voltage range: 4 to 15 V
- Voltage gain: 200 V/mV
- Unity gain bandwidth: 3 MHz

Figure 11.29 shows the pinout for these operational amplifiers.

11.5 Suggested Projects

1. Create a configuration that makes a robot follow a dark stripe on the ground.
2. Design a robot that uses two LDRs to find a light source such as a lamp or a candle. Use voltage comparators for this project.
3. Design a system, using the window comparator, to add intelligence to a robot that looks for a lamp, but make it "afraid" of excessive light, keeping a "safe" distance from it.
4. Create an arm that can be controlled by potentiometers, using other potentiometers as position sensors.
5. Add a temperature sensor to a robot that allows it to separate hot objects from cold objects.

11.6 Review Questions

1. How many inputs does an operational amplifier have?
2. After amplification, what happens to the phase of a signal that is applied to the inverting input of an operational amplifier?
3. What is the output of a comparator if the signal delivered to the noninverting input is higher than the reference voltage?

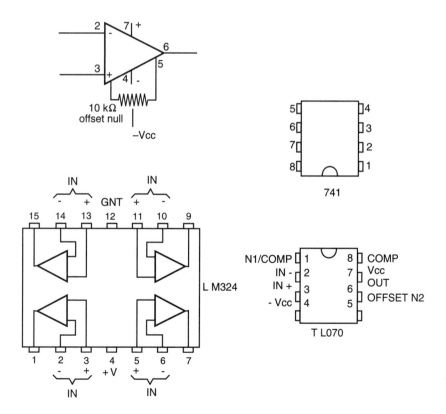

Figure 11.29 Pinouts of some opamps.

4. How many comparators are used in a window comparator?
5. How many comparators exist in an LM324 IC?
6. What is a dual-voltage power supply (e.g., as needed to power comparators and operational amplifiers)?

12

Remote Controls and Remote Sensing

12.1 Purpose

The purpose of this chapter is to present remote control blocks that are suitable for projects in robotics and mechatronics. Many technologies can be used, and the choice depends on the specific applications. In addition to basic theory, this part offers blocks of different types, ranges, and applications, allowing the reader to make the right choice for an application.

12.2 Theory

Many simple circuits exist that are suitable for remote control and remote sensing. It is important to know something about these circuits in terms of basic operational theory and possible applications.

A remote control can be defined as a system that allows an operator to send control signals to a robot or mechatronic device. The simplest type can communicate via a long wire between a keyboard and a device. But the more important ones use no physical medium between the controller and the controlled unit. In this group of applications, we include remote sensors that employ some kind of transmitter and receiver, sending information from sensors to a remote circuit.

12.2.1 Wires

The simplest remote control system is one that uses wires to connect a keyboard or a control unit to the device to be controlled, as shown in Figure 12.1. In particular, this kind of system is suitable for applications in which the controlled unit is in a fixed location. A fixed robotic arm and an elevator are examples of devices in which this kind of control is applied. In some cases, a robot can also be controlled by wires if it is intended to move only in a restricted area.

12.2.2 Light

Light beams can be used to send signals from a control unit to a controlled device. The light beam's advantages include low power drain and ease of manufacture. Another advantage to the use of light beams is that they are not subject to electromagnetic interference (EMI). The biggest disadvantages are that the beam must be

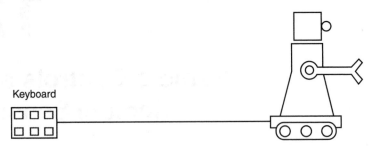

Figure 12.1   Simplest remote control.

aimed directly at the sensor, and obstacles between the transmitter and sensor can affect the operation of the device. The simplest light beam remote control can employ a flashlight as transmitter, as we will show in a block.

12.3   Infrared

Most domestic appliances (TVs, videocassette players, DVDs, etc.) use remote control systems that are based on infrared radiation (IR). This system is formed by a infrared transmitter and a receiver. The infrared radiation is modulated by the signal to be transmitted. In the receiver, the signal is decoded and sent to the controlled circuit.

The advantages of this system are that IR is invisible, has a good working range, and is not subject to EMI. Low-cost emitters and transmitters can be used in systems with several channels of information and at ranges up to 50 m.

The disadvantages are the same as with a light beam system: the IR beam must be aimed directly at the sensor, and obstacles can block the operation of the system.

12.3.1   Sound

Sound waves can also be used to send control signals to a robot or mechatronic device. A tone generator, or even a person's whistle, can be used to turn a relay on and off and stop or start a robot. By emitting and sensing sounds at various frequencies, multiple functions can be controlled as shown in Figure 12.2.

For simple purposes, a robot that recognizes individual musical notes is quite attractive. A more complex project can include voice recognition, but this off the track of our basic purpose, which is to describe simple blocks for the student or beginner.

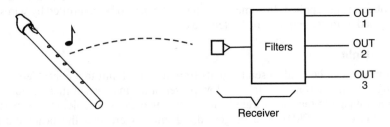

Figure 12.2   A flute used as a remote control transmitter.

The advantage to the use of sound waves is that they are very simple to produce (in some cases, no electronic devices are needed). The disadvantage is that ambient noise can interfere with the system.

Many systems can use ultrasonic, inaudible sounds. Transducers suitable for these applications are easy to find.

12.3.2 Radio

Radio waves have special advantages when used with robotic and mechatronic devices. Simple circuits can send signals to distances up to 100 m, with the advantage that they can pass around or through solid obstacles such as walls, wood panels, glass, and other materials. An additional advantage is that the signals need not be directed at the receiver.

In the basic system, we have a radio transmitter that sends coded signals to a receiver as shown in Figure 12.3. The frequency of the system is an important detail to be considered when designing a radio-controlled project. The simplest systems can use the medium wave (MW) band, using portable radios as receivers. Others might use the 40–50 MHz FM band, which also allows the use of simple portable radio receivers. Still others can operate in the ultrahigh frequency (UHF) range; using hybrid modules, we can reach 300 to 400 MHz.

The disadvantage of using radio waves is the widespread presence of electromagnetic interference (EMI) in the ambient. Depending on the application, EMI levels can be substantial enough to rule out the use of this kind of system.

12.3.3 Electromagnetic Interference

Domestic appliances, industrial and medical machines, automobiles, and many other devices produce radio waves as consequence of their natural operation. A simple electric motor's brushes, in switching among windings on the armature, produce substantial electromagnetic waves that can span the radio spectrum.

Although filters and other devices are used in appliances to reduce or eliminate appliance-generated radio waves, they often are not efficient enough, and EMI can appear in remote control systems and radio receivers.

To demonstrate how this works, tune a small MW radio to a frequency between stations and place it near a computer or an electric motor. The result will be as shown in Figure 12.4.

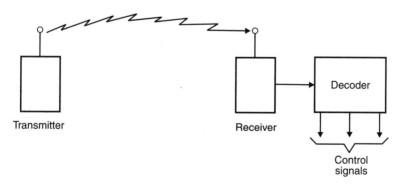

Figure 12.3   Simple radio control.

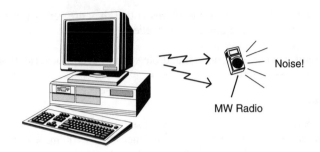

Figure 12.4   Noise produced by a computer.

In cases where EMI levels are high, some solutions are available when using a remote control system:

- Install filters in the receiver or in the emitting device if possible.
- Use as high a frequency as possible to reduce EMI.
- Choose a different kind of system (IR or sound).

12.4   Choice and Use of a Remote Control

The choice of a remote control for an application in robotics and mechatronics depends on several factors.

- First, determine if the control can be hard wired or if the application demands wireless control.
- If wireless, choose the appropriate medium: sound, radio, visible light, or IR.
- Analyze the advantages and disadvantages of each system for each specific application, taking into account
  -The presence of obstacles
  -The required range
  -The existence of potential interference sources
  -System efficiency
- Once you have selected the appropriate system, find a circuit suitable for the specific application.

12.5   Blocks

In the following paragraphs, we present blocks that can be used as a complete remote control system or wired to other blocks to as part of a more complex system. Many of the blocks need some additional parts, sometimes including blocks suggested in other parts of this book. Occasionally, changes must be made in component values to achieve the best performance. Is up to the reader make the necessary adaptations for the specific tasks on a case-by-case basis.

Block 185   Remote Control Using Wires

Many of the blocks shown in Chapter 2, Motion Controls, can be used as remote control by wires. As an example, we use Block 5 in this application.

In Figure 12.5, the block is used to control two loads (relays, motors, or other circuits) from a remote keyboard. The remote station uses a battery to power the circuits in the controlled unit and two switches. One of the switches selects the function (circuit 1 or circuit 2), and the other activates the selected circuit.

Some other blocks can be used as remote controls simply by connecting them to relays, solenoids, or other loads rather than to motors. These include the following:

- Block 1: Two wires on-off
- Block 2: Two wires reversing
- Block 4: Three wires, selecting one of two loads
- Block 6: Two wires, ac control
- Block 20: Two wires, two loads
- Block 23: Two wires, series/parallel

Block 186   Multi-wire Control

This circuit can be used to control many functions of a robot or mechatronic device using a flat cable and a keyboard. The circuit given in Figure 12.6 uses three momentary contact switches (pushbuttons) but, depending on the function, the switches can be SPST or another type, and the number can be increased.

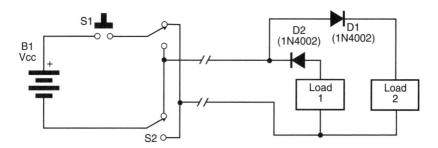

Figure 12.5   Block 185: remote control using wires.

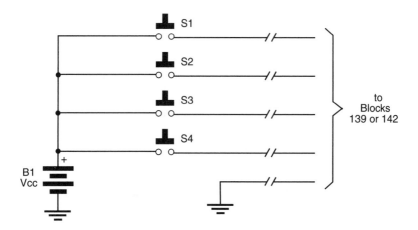

Figure 12.6   Block 186: multi-wire control.

The basic application shows relays as the load, with the contacts controlling the desired function. However, other blocks can be controlled, and the controller can even drive a circuit if the battery used in the transmitter is powerful enough. One advantage of the use of this kind of control is that two or more functions can be activated at the same time.

The blocks driven by this circuit depend on the application. If you need to power a circuit for a particular time interval by pressing a momentary contact switch, you can use Block 139. If you want a bistable control, you can use the circuits shown in Block 142.

The maximum length of the cable is limited by two factors:

1. *Power losses in the wires.* If power circuits are driven directly from the transmitter, unit the length of the cable is limited to about 10 m.
2. *Sensitivity to noise.* If high-impedance circuits are controlled directly from the keyboard, noise can induce erratic operation.

Note: Many other blocks described in previous parts of this book can be used as remote control units employing wires. In particular, consider the blocks used for the control of motors (motion), relays, on-off sensors, and for debouncing.

Block 187   Sequential Control Using Wires

The block shown in Figure 12.7 provides the robotics and mechatronics designer with many application possibilities. At each touch of the control pushbutton, one of a programmed series of functions is activated. The pulses can also be produced automatically by a relay connected to some of the blocks shown in other parts of this book.

The circuit can handle two to ten functions, as the 4017 IC has 10 independent outputs. At each touch, one output goes to the high logic level (positive supply voltage) and remains in this state until the next pulse arrives. The next pulse makes

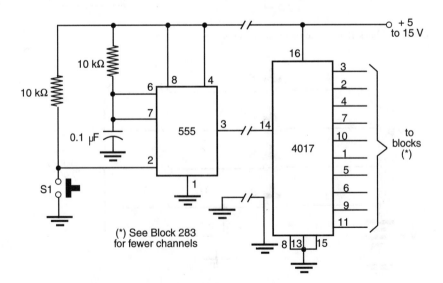

Figure 12.7   Block 187: sequential control using wires.

the output return to the low logic level and, at the same time, the next output is activated.

To program the number of outputs, you just need to connect the output that follows the last one used to the desired RESET input. For instance, if you want four outputs, connect the fifth one to the reset as shown in Figure 12.8.

The outputs of the 4017 can control any of the blocks suggested in Block 167 or even Block 168 if you want to turn off a load when the output goes to the high logic level. The block can be powered from 5 to 12 V supplies, and the maximum length of the wires is 100 m. A 470 Ω resistor can be added in parallel with the line to reduce the risk of interference.

Block 188   Matrix Control

Many functions can be activated by hard-wired controls if a matrix decoder is used. The block shown in Figure 12.9 gives an example of a simple matrix using diodes. In this example, the keyboard in the control unit has three pushbuttons. The signals are sent to the controlled unit, in which a card with a program matrix is installed (the reader can use different cards if necessary).

When one pushbutton is pressed, the corresponding diodes conduct, and in the output we have one to four control signals. For the example shown in the figure, the following table gives the outputs as function of the inputs:

Switch	OUT 1	OUT2	OUT3	OUT4
S1	1	0	0	1
S2	0	1	1	0
S3	1	0	0	0

The size of the matrix depends on the number of switches and the number of outputs. All the control blocks described in the previous chapters, such as the ones suggested in Block 167, can be controlled by this circuit. The diodes can be 1N4148, 1N914, or 1N4002 units if you want to directly power loads up to 1 A. An interest-

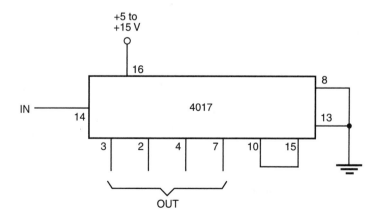

Figure 12.8   Block 188: matrix control.

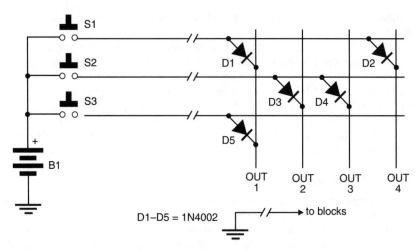

Figure 12.9   Block 188: matrix control.

ing application for this block is in combination with Block 187, which allows the control of more than one load by each pulse of the control unit.

Block 189   Tone Generator

Tone generators are used in many types of remote control systems. In wired, radio, or IR remote controls, tone modulation improves the liability of the circuit, as we will see in the following blocks. Tone generator I, illustrated in Figure 12.10, will operate with many of them.

This circuit produces a tone in the range from 100 to 10,000 Hz. This tone is advantageous in controlling remote units, since it can be transmitted more easily than a dc control signal. The circuit can be powered from 4 AA cells or a 9 V battery and is

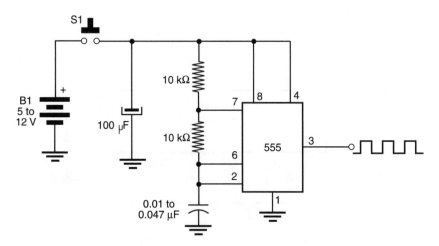

Figure 12.10   Block 189: tone generator.

activated when S1 is pressed. S1 can be replaced by relays or other sensors as appropriate for the application.

Block 190   AC Switch (I)

The simple circuit shown in Figure 12.11 drives a relay, turning it on when an audio
tone is applied to the input. This circuit can be used with the previous block in a
wired remote control. The advantage in the use low-frequency signals is that the
range can be extended to more than half a mile with common wires.

The circuit can be powered from 6 to 12 V, depending on the relay. The capacitor
is installed to avoid interference that can make the relay turn on when you don't
want it to. Resistor R1 must be chosen according to the amplitude of the signal. Start
with a 10 kΩ device and move down toward 1 kΩ until you find the best performance for the block. This circuit needs at least 1 Vpp to be activated.

This circuit is untuned, meaning that it cannot recognize any specific frequency. It
switches the relay on when it receives a tone at any frequency.

Block 191   AC Switch (II)

The block in Figure 12.12 uses two complementary transistors. This circuit can operate with very weak tones coming in from the remote control unit or a sensor.

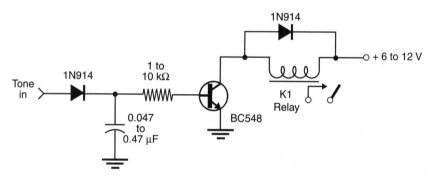

Figure 12.11   Block 190: ac switch I.

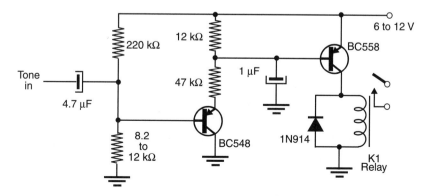

Figure 12.12   Block 191: ac switch II.

The circuit can drive a relay or other control circuit, depending on the application. Replacing the PNP transistor with a BD136, for instance, allows you to control loads up to 500 mA with this block. The power supply voltage depends basically on the relay or the load to be controlled.

This is also an untuned circuit, triggering with tones of any frequency. Capacitor C1 will be valued at between 1 and 10 µF, according to the frequency of the tone. It is installed to avoid problems caused by contact vibration in the relay.

Block 192   PLL Tone Decoder

Phase locked loops (PPLs) are circuits than can be tuned to recognize a signal of predetermined frequency. They are very useful as tone decoders in multichannel remote control systems. This simple block uses one of the more popular ICs designed for this task: the LM567. The basic block of a tone decoder that can operate with signals in the range between 0.1 Hz and 500 kHz is shown in Figure 12.13.

In this basic configuration, when the circuit, using the trimmer potentiometer P1, is tuned to the proper frequency, the output goes to the low logic level, saturating the transistor. At this point, the relay turns on, switching an external load.

The 567 is very sensitive, requiring signals of only 20 to 100 mV to be triggered. It is also very selective, rejecting signals whose frequencies differ only slightly from the tuned frequency.

This block can be used to detect the tone sent by a remote transmitter such as the one shown in Block 189, even when using very long wires (up to a mile). One important point is that the 567 must be powered with voltages in the range of 4.75 to 9 V. If you need to drive a 12 V relay, the output stage can be powered from a separate voltage source.

This block can also be used to trigger other blocks shown in this book, such as the ones that power on loads when the input goes to the low logic level. With the values recommended in the diagram for C1 and P1, the circuit can be adjusted within a frequency range of 500 to 5000 Hz, the same as generated by Block 189. You can also tune the transmitter by replacing one of the resistors by a 100 kΩ trimmer potentiometer in series with a 2.2k Ω resistor.

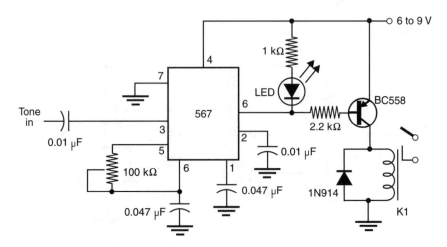

Figure 12.13   Block 192: ac switch II.

Block 193   Multi-tone Transmitter

Using the LM567 IC as in Block 192, it is easy to design a multichannel remote control using wires (or radio or IR, as we will see in the next blocks). The first step is to design a block to generate multiple tones when the corresponding switch is activated on a keyboard. An important block for this task is given in Figure 12.14.

In this basic version, the block can produce three different tones that are determined by adjusting the trimmer potentiometers. When S1 is pressed, the 555 is powered on by the corresponding diode, and the R/C network that determines the frequency is connected to the circuit. This circuit can be powered from 6 to 9 V supplies.

Block 194   Multichannel Tone Decoder

This block is designed to operated with multi-tone remote control systems, and the first application is as a receiver for the transmitter shown in Block 193. The version for a three-channel system is given in Figure 12.15.

Remember that the LM567 is powered from 4.75 to 9 V supplies, but the relay can be driven from a higher voltage source. Another point to consider when using this circuit is that the operating frequency of each channel must be carefully chosen. The frequency of each channel must not be a multiple or harmonic of any other channel. If you adjust one channel to 1 kHz and the other to 2 kHz, for instance, you may have problems, since the PLL will find it difficult to separate harmonics from the fundamental. Suitable frequencies are 1 kHz, 1.2 kHz, 1.4 kHz, 2.1 kHz, etc.

This fact severely limits the number of channels you can use with this system. As general rule, you need to limit the number of channels to six or seven.

The output of each tone decoder can be connected to a relay driver, as shown in Block 192, or any control block that turns the load on when the output of the PLL goes to the low logic level.

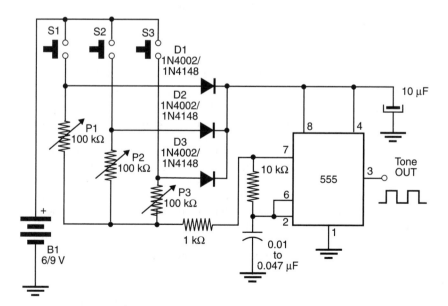

Figure 12.14   Block 193: multi-tone transmitter.

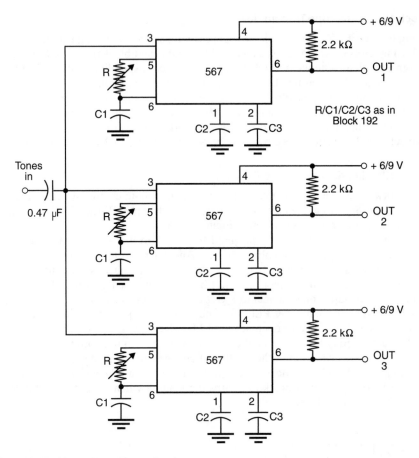

Figure 12.15   Block 194: multi-tone decoder.

Block 195   Remote Control Using the AC Power Line, Transmitter

The wires of the ac power line can be used to carry control signals from one part of a building to others. The range is determined by many factors such as the presence of noise, inductive loads, etc. As a general rule, if the transmitter and the receiver are connected to the same line, the range will be up to 20 m inside a building. This makes the block shown in Figure 12.16 suitable for many indoor applications. For best results, the operating frequency must be in the range of 40 to 100 kHz.

The block shown here is the transmitter. It is basically a power stage that drives the ac power line by two capacitors. These capacitors isolate the line from the circuit, as they represent high impedance for the low-frequency voltage of the ac power line. The capacitors must be polyester types with a working voltage of at least 250 V.

The block is driven by Block 189 in a one-channel system or by Block 193 in a multichannel system. The transistor must be mounted on a heat sink.

Block 196   Remote Control Using the AC Power Line, Receiver

Since the signals produced by this system are strong, the receiver can be very simple; as shown by Figure 12.17, it is an isolation transformer that picks up the high-

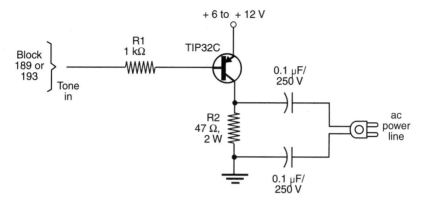

Figure 12.16   Block 195: ac power line transmitter.

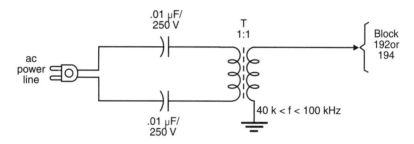

Figure 12.17   Block 196: remote control using the ac power line, receiver.

frequency signals from the ac power line. The transformer is formed by 100 + 100 turns of 32 AWG enameled wire on a ferrite rod. The rod is 4 cm long and has a diameter between 0.5 and 1 cm.

As in the previous project, the capacitors must be polyester types, with at least 250 V working voltage. This block can be connected to Block 192 in a one-channel system or to Block 194 for a multichannel system.

Block 197   Flashlight Remote Control

A flashlight can be used in a simple remote control to switch a load on for a particular time interval. A block that is suitable for this task is shown in Figure 12.18.

The on-time the relay is determined by the RC time constant. It can be calculated by the following formula:

$$t = 1.1 \times R \times C$$

where   t = time in seconds
           C = capacitance in Farads
           R = resistance in ohms

For a 1 MΩ resistor and a 100 μF capacitor, the time is about 2 min.

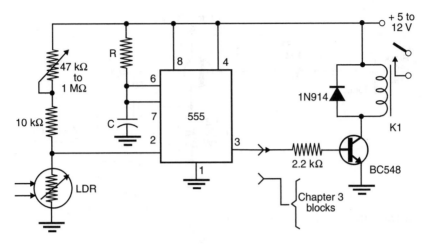

Figure 12.18 Block 197: flashlight receiver.

A short light pulse is sufficient to turn on the load. You can increase system sensitivity and directivity by mounting the LDR inside a tube that is fitted with a lens.

This block can be used to control other blocks as the ones shown in part 3. Many blocks that use LDRs as sensors can also be employed in remote control systems where the transmitter is a flashlight.

Block 198   Infrared Transmitter

The block shown in Figure 12.19 can be used to modulate two IR LEDs from a tone generator such as those described in Blocks 189 and 193. In the first case, we have a single-channel system, and in the second case a multichannel one. The transistor must be mounted on heat sink and, since the circuit is on only during a short activation time, the power supply can be common cells or a 9 V battery. The circuit must be powered on only when the keyboard is activated; when the IC is set with the low output, the transistor drains a high current.

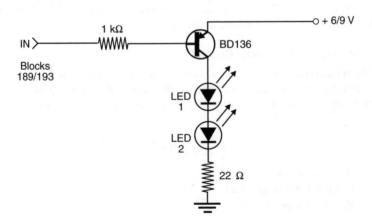

Figure 12.19 Block 198: infrared transmitter.

Block 199   Infrared Receiver

This block can be used as a receiver for the signals produced by the previous block. The circuit shown in Figure 12.20 uses a photodiode or phototransistor as a sensor, since this kind of device responds very quickly and can operate with light or IR modulated by a high-frequency signal.

Feedback resistor R1 determines the gain of the circuit and, depending on the desired range, its value is between 1 and 10 MΩ. The trimmer potentiometer adjusts sensitivity.

To get better performance, the sensor (photodiode or phototransistor) can be mounted inside an opaque tube with a lens. This circuit can drive Block 192 or Block 194 to act as a single-channel or multichannel remote control system, respectively.

Block 200   Sound/Ultrasonic Transmitter

The circuit shown in Figure 12.21 can be used to produce a sound in the audible range or, depending on the transducer, in the ultrasonic range (above 18 kHz). The

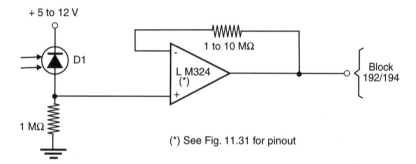

Figure 12.20   Block 199: infrared receiver.

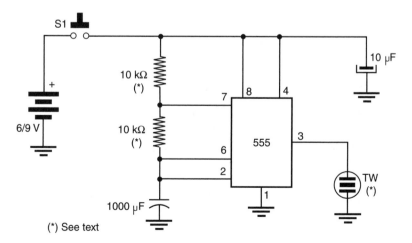

Figure 12.21   Block 200: sound/ultrasonic transmitter.

most economical solution in the audible range is to use a small piezoelectric transducer (high-impedance) driven directly from the output of the 555. The frequency of this tone is determined by R1 and R2. R1 can be replaced with a 100 kΩ trimmer potentiometer in series with a 2.2 kΩ resistor if adjustment is required.

For operation in the ultrasonic range, a piezoelectric tweeter can be used. Remove the small transformer inside the tweeter, since it is used to convert the high impedance of the piezo element to low impedance, and drive the transducer directly from the output of the IC. These tweeters can produce ultrasonic sounds in a range up to 25 kHz with good performance.

The basic version for a single-tone transmitter is given in this block, but the oscillator stage can be adapted easily for the configuration shown in Block 193 in a multi-tone (or multichannel) system.

To drive a low-impedance tweeter directly (without having to remove the transformer), use the configuration of Block 198, replacing the two IR LEDs and the 22 Ω resistor with the transducer (tweeter).

Block 201   Low-Impedance Sound/Ultrasonic Receiver

Low-impedance transducers can be used to receive sound and ultrasonic waves. A piezoelectric tweeter or even a small loudspeaker can be used as the receiver with the block shown Figure 12.22, which acts as a preamplifier with a high-impedance output.

This block can be connected directly to the input of the tone decoders of Blocks 192 and 194. This circuit can be used with magnetic transducers having impedances ranging between 4 and 200 Ω.

Block 202   Audible Receiver for Electret Microphone

The block shown by Figure 12.23 is suitable for operation with audible tones. This block uses a two-terminal electret microphone, and the gain is set by the feedback resistor. The diagram suggests the range of values that are appropriate for this component. The block can also be used to drive decoders such as the ones shown in Blocks 192 and 194, and even Block 190 for a single tone.

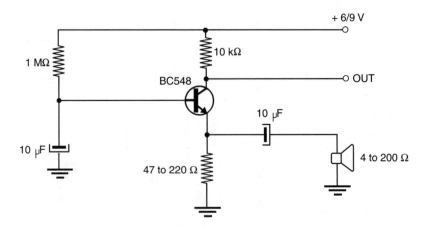

Figure 12.22   Block 201: low-impedance receiver.

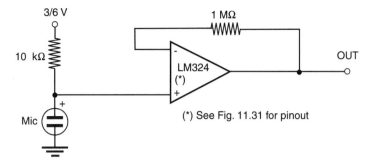

Figure 12.23   Block 202: audible receiver.

Block 203   30 to 100 MHz Transmitter Block

The circuit shown in Figure 12.24 is a small VHF transmitter that can send signals over distances up to 50 m. The range depends on the sensitivity of the receiver, the presence of obstacles between transmitter and receiver, ambient noise, and the characteristics of the antenna. The antenna used here is a piece of rigid wire, 30 to 80 cm long.

The coil is wound with 28 or 26 AWG wire on a coreless form 1 cm in diameter. The relationship between the number of turns and the frequency is given by the following table:

Frequency Range	No. of turns
30 to 50 MHz	10
50 to 70 MHz	6
70 to 90 MHz	4
90 to 100 MHz	3

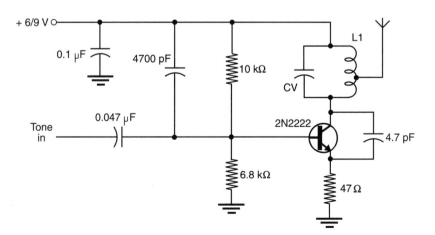

Figure 12.24   Block 203: 30 to 100 MHz transmitter.

All capacitors must be ceramic types, and the modulation signal is a tone. CV adjusts the operating frequency.

Block 204   Using a Small FM Radio as Remote Control Receiver

If the circuit shown in Block 203 is adjusted to operate in the FM range, a portable FM radio can be used as a receiver as shown in Figure 12.25. It can be the source of a signal for the tone decoders of Blocks 190, 192, and 194, deriving the output via a phone jack, using a 0.01 µF capacitor in the line if necessary. This circuit is appropriate if the modulation tone is in the range of 200 to 10,000 Hz. Any trimmer capacitor with capacitances ranging between 20 and 40 pF (maximum) can be used.

Block 205   30 to 100 MHz Super-Regenerative Receiver

Figure 12.26 shows a super-regenerative receiver for the VHF band, suitable for applications in short-range remote control. This circuit can operate in the range between 30 and 100 MHz, depending on the coils as given in the table of Block 203.

XRF is a 100 µH micro-choke, but you can substitute a home-made unit; it can be built by winding 40 to 60 turns of 32 to 26 AWG enameled wire on a 100 kΩ × 1/2 W resistor.

The antenna is a piece of rigid wire 30 to 80 cm long, and P1 adjusts the operating frequency. To adjust the circuit, connect its output to the input of an audio amplifier, then tune CV and adjust P1 to receive the tone produced by the transmitter.

Block 206   Coder Using the MC145026

The block shown in Figure 12.27 uses the Motorola MC145026 IC, which can encode nine bits of information and transmit this information serially. This basic block can be used to send one of 19,683 codes, making it virtually impossible for another circuit to decode the same signal. The code is set in the transmitter, and the receiver must be set to recognize the same code.

The block can be powered from 4.5 to 18 V supplies. The same block can be used to send the serial information through wires or via an infrared transmitter as shown in other blocks in this chapter. The timing capacitors and resistor are critical elements in this project, and the indicated values must be observed.

Block 207   Decoder Using the MC145026

Figure 12.28 shows the decoder block for the circuit shown in the previous block. This block can be wired directly to the output of blocks that use IR or sound. More

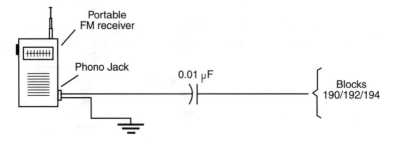

Figure 12.25   Block 204: small FM radio used as a radio control receiver.

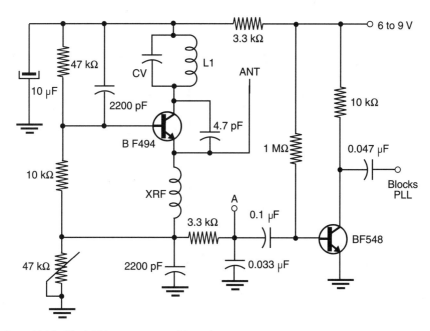

Figure 12.26   Block 205: super-regenerative receiver.

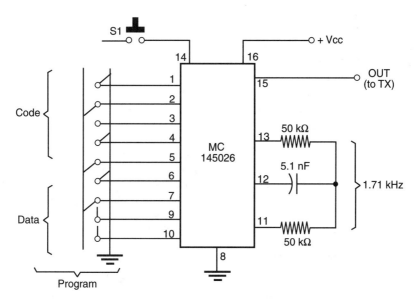

Figure 12.27   Block 206: coder using the MC145026.

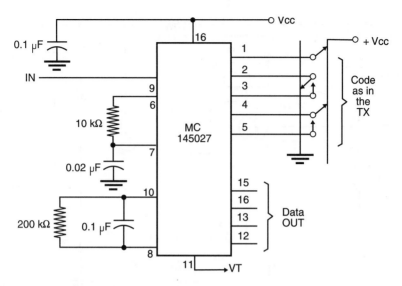

Figure 12.28   Block 207: receiver for Block 206.

information about the use of this block, and the previous one, in remote control systems can be found at Motorola's web site (http://sps.motorola.com). The MC145026/27 pair can also be used in multichannel systems.

12.6   Additional Information

The LM567 is a very important component for the robotics and mechatronics designer. Below, we give some of the characteristics of this IC along with its block diagram (Figure 12.29).

Characteristics
- Power supply range: 4.75 to 9 V
- Frequency range: 0.01 Hz to 500 kHz
- Adjustment range: 20:1
- Sensitivity (smallest detectable input voltage): 20 mVrms (typical)
- Input resistance: 20 kΩ (typical)
- Maximum current sinking: 100 mA

Figure 12.30 gives some basic applications for this circuit. In 12.30a, we have the application of a tone decoder driving a relay or other load. In 12.30b, we have an FM discriminator. This circuit can extract the modulation from a modulated signal. With this circuit, it is possible to use a high-frequency carrier (e.g., 100 kHz) to carry low frequency tones (1 kHz, 1.5 kHz, etc.). The low-frequency tones appear in the output of this circuit and can be used to drive the tone decoders.

12.7   Suggested Projects

Combining all the blocks that have been described in this book, the reader can conceive of many interesting robots and mechatronics devices. Try to combine these blocks to create the following configurations:

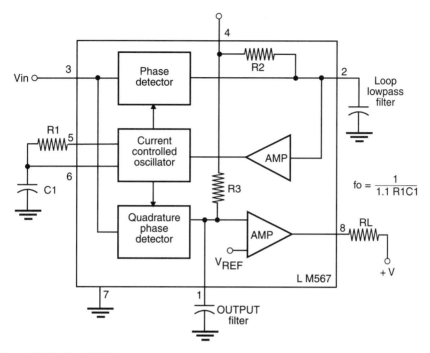

Figure 12.29 The LM567.

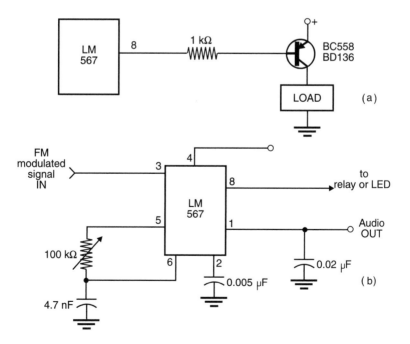

Figure 12.30 Applications for the LM567.

1. Design a remote controlled robot that can be driven from a keyboard but that also has "intelligent" functions using sensors and other resources.
2. Create a robot that can send you information when certain signals are received by its sensor. It then waits for a confirmation using a remote control unit before it proceeds with another operation.
3. Design a simple, remote controlled mechatronic arm.
4. Create an "intelligent elevator" that also has some remote controlled functions.
5. Design a robot that performs multiple tasks, and use a diode matrix to program it.

12.8   Review Questions

1. What is the difference between a remote control and a wireless control?
2. What are the main disadvantages of a wired remote control?
3. What are the disadvantages of a sound-operated remote control?
4. How can EMI affect remote control systems that use radio waves?
5. What is a PLL?
6. What is a hybrid module?
7. Why are frequencies in the UHF range more suitable for applications in short-distance remote control?

13

Logic Blocks

13.1 Purpose

Logic blocks can add intelligence to a robot or a mechatronic device. This chapter shows the reader how some simple blocks incorporating digital logic can be used to program a sequence of operations or to add decision-making capabilities to a robot. The blocks are very simple but can be wired together to perform complex functions. They can be the starting point for very complex designs using microcontrollers or microprocessors. These blocks can also be seen as complementary to complex projects that include logic, since they perform simple functions that do not require complex circuitry.

13.2 Theory

Digital circuits operate with only zeros and ones (binary digits, or *bits*): the zero or low logic level corresponds to 0 V, or the ground potential, and the one or high logic level corresponds to a positive voltage—normally, the power supply voltage. These circuits cannot operate with intermediate voltages. The logic blocks process digital signals from which they generate new digital signals according to preset rules.

Combining these blocks in different ways, it is possible to perform complex mathematic or logic operations as is performed by computers. Even complex computers are made from simple basic blocks called *gates, inverters, counters, memories, flip-flops,* and so on.

Simple logic blocks can allow your robotics or mechatronics project to make some of its own decisions according to its programming. The basic blocks described here use easy-to-find components, and many of them have complex and sophisticated versions in the form of integrated circuits.

For didactic and experimental purposes, simple blocks can be much more important than complex ones, since they more easily understood and used. With an understanding of how these basic blocks operate, it will be easier for the student or beginner to step up to more complex blocks such as microprocessors, digital signal processors (DSPs), and microcontrollers.

13.2.1 Gates, Inverters, and Buffers

Gates are used to implement simple logic functions such as OR, NOR, AND, NAND, and NOR. Other simple blocks that can be included in this group are inverters and buffers. To represent what logic blocks such as gates and inverters do, we can use *truth tables*. These tables indicate what happens to the output when all possible signals are applied to the input of a particular device.

Buffer

Input	Output
0	0
1	1

The buffer is used as a "digital" amplifier, driving loads that are not supported by other common logic functions.

Inverter

Input	Output
0	1
1	0

AND Gate

The output of a two-input AND gate goes to the high logic level (1) when input 1 AND input 2 are set to the high logic level.

Input 1	Input 2	Output
0	0	0
0	1	0
1	0	0
1	1	1

OR Gate

The output of a two-input OR gate is at the high logic level when input 1 OR input 2 is at the high logic level.

Input 1	Input 2	Output
0	0	0
0	1	1
1	0	1
1	1	1

NAND Gate

The output of a NAND gate is at the high logic level when input 1 AND input 2 are NOT at the high logic level.

Input 1	Input 2	Output
0	0	1
0	1	1
1	0	1
1	1	0

NOR Gate

The output of a two-input NOR gate is at the high logic level when input 1 OR input 2 is NOT at the high logic level.

Input 1	Input 2	Output
0	0	1
0	1	0
1	0	0
1	1	0

Exclusive OR

The output of a two-input exclusive OR gate is at the high logic level when input 1 OR input 2 is at the high logic level, but NOT both.

Input 1	Input 2	Output
0	0	0
0	1	1
1	0	1
1	1	0

13.2.2 Monostable/Bistable

A monostable can be defined as *a circuit with only one stable state.* Monostable blocks are very important in robotics and mechatronics projects, since they can be used to control operational timing or employed simply as debouncers. In previous chapters, we observed how some of these blocks can be used for critical functions.

Once triggered, the monostable is forced into an unstable state, at which it remains only for a specified time. After this interval, it returns to its initial state. Figure 13.1 shows a block representing a monostable.

The time period during which the monostable remains in the unstable state is normally determined by an RC network. In robotics and mechatronics, monostables can be used as timers to govern how long a circuit is energized after it is triggered.

In contrast, the bistable circuit has two stable states, as shown in Figure 13.2. Let us assume that the circuit is in a state at which output Q1 is high, and Q2 is low. When triggered, the circuit changes its state: output Q1 goes to low, and Q2 goes to high, and both remain this way until a new pulse is applied to the circuit. When the new pulse comes, the circuit again changes its state; Q1 reverts to the high state, and Q2 to the low state.

The bistable is also called a *flip-flop,* and it can be used as a single-bit memory, a divide-by-two circuit, or to trigger loads on and off using single-pulse commands. Bistables and monostables can be found in the form of integrated circuits.

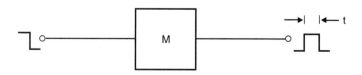

Figure 13.1 A monostable block.

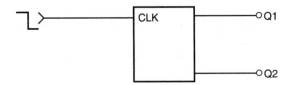

Figure 13.2   A bistable block.

13.2.3 Counters and Decoders

Blocks based on flip-flops can be used to count pulses or to divide the frequency of a signal. The number of pulses counted by a counter can be stored and, in some types, decoded by other blocks *(decoders)*. Counters and decoders can be found in the form of ICs of both TTL and the CMOS families. Many important blocks are based on counters and decoders, since these devices allow us to program a circuit's logic operations or control a circuit in automated operations.

13.2.4 Voltage-Controlled Oscillators

Voltage-controlled oscillators (VCOs) are circuits that produce a signal, the frequency of which depends on the voltage applied to the input. They can be used as logic blocks in projects that do not include sensors, for sending information to a remote terminal, and also as blocks of artificial intelligence. This will be described in subsequent text.

Many simple VCOs configurations can be employed, and there also are some integrated circuits that incorporate all of the elements needed to perform this function with the help of few external components. We will describe one important VCO block that is suitable for applications in robotics and mechatronics.

13.2.5 Memory

Memories are blocks used to store information. The information can be a single bit (0 or 1) in a digital memory or a voltage in an analog memory. Digital memories are very popular, since they can be obtained easily in the form of integrated circuits. These memories can store a huge number of bits, with capacities as high as many megabytes.

Memories can be used to store information such as the sequence of operations of a robot or a mechatronic devices (its *program*). This program tells it how to perform all of the tasks that we want it to do.

The simplest memories can be made with diodes, but other types are compatible with TTL and CMOS logic and can be used in practical projects. The blocks provided herein will include examples of both types of digital memory, and also of analog memory.

13.3 How to Use Logic Blocks

When using logic blocks, it is important to know what kind of input signals they need and what kind of signal we can get from their outputs. TTL blocks must be powered from 5 V supplies and can only output a few milliamperes. They typically can deliver 1.6 mA and drain 16 mA at their outputs. CMOS blocks can drain or sink 2.2 mA at the outputs when powered from a 10 V power supply. They can be powered from voltages from 3 to 15 V, but operation between 3 and 5 V is sometimes not recommended.

13.4 Blocks

The following are blocks that use logic functions, performed in some cases by discrete components or ICs. Many of the circuits are developed from blocks shown in previous chapters. Some are intended to be used as shown, but others can be upgraded to perform much more complex functions than the ones suggested here. It is up to reader make the necessary modifications to each block to solve a particular application problem.

Block 208   NOR Gate Using a Transistor

Simple logic functions can be implemented with a common bipolar transistor as shown in the block of Figure 13.3. The output of this block goes to the low logic level when IN1, IN2, or IN3 goes to the high logic level.

This block can directly control a load if you use an appropriate transistor (see Chapter 3 for suggestions about high-power blocks). Using the BD135, you can control loads of up to 500 mA. Replacing the load by a 1 to 2.2 kΩ resistor, and using the BC548, the block can be used as a logic gate to drive other blocks. This circuit can be powered from 5 to 12 V sources.

Block 209   NAND Gate Using a Transistor

Figure 13.4 shows how to use an NPN bipolar transistor as a NOR gate. Using a BD135 transistor, you can directly drive a load of up to 500 mA. The output of this circuit goes to the LOW logic level when IN1, IN2, and IN3 go to the high logic level.

If you use a BD135, the circuit can directly drive loads up to 500 mA. If you replace the load with a 1 to 2.2 kΩ resistor, and the transistor with a BC548 device, the circuit can be used to drive other blocks.

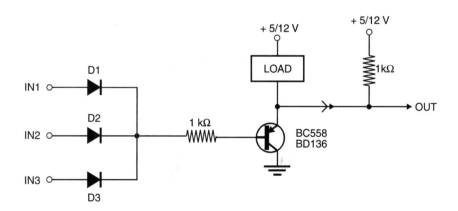

Figure 13.3   Block 208: NOR gate using a bipolar transistor.

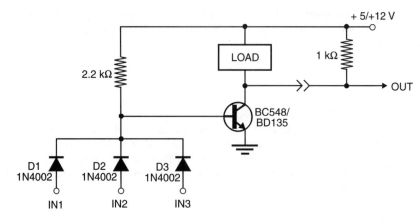

Figure 13.4   Block 209: NAND gate.

Block 210   Logic Inverter Using a Transistor

A simple block using a bipolar NPN transistor for a logic inverter or NOT logic function is shown in Figure 13.5. In this block, the output goes to the low logic level when the input is high. The circuit can directly drive loads or other blocks as in the previous blocks.

Block 211   Basic Monostable Using the 555 IC

Monostable or timing functions are very important when designing robots or mechatronics devices. The block shown in Figure 13.6 is probably the most important of the monostables, since it uses the popular 555 IC.

In this block when the input is momentarily set at ground potential, the output of the 555 goes to the high logic level for a time interval determined by R and C. This time interval can be calculated by the following formula:

$$t = 1.1 \times R \times C$$

where   t = time in seconds
R = resistance in ohms
C = capacitance in farads

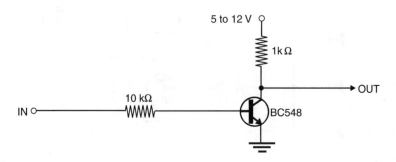

Figure 13.5   Block 210: logic inverter.

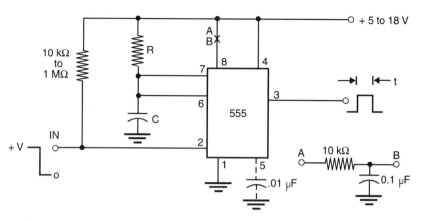

Figure 13.6   Block 211: basic monostable using the 555 IC.

This block is suitable for time intervals from a fraction of second to more than an hour. Longer time intervals are not recommended, because losses in the capacitor can affect the performance of the circuit.

The same configuration can employ the TLC7555, which is the CMOS version of the 555 bipolar. This block can directly drive loads of up to 200 mA or more powerful blocks as the suggested in the figure.

In some applications, you must make sure that the circuit is reset when the power is switched on. This can be done using the network format of a resistor and a capacitor placed between points A and B.

Block 212   Sequential Monostable

Monostables can be triggered in sequence, producing two or more timing intervals. The 555 ICs are wired in series as shown in Figure 13.7.

The first monostable is triggered when its input is momentarily set to ground potential. When the first timing interval is over, the second monoestable is triggered, and its output remains high for the length of the second time interval.

The sequence of the monostable can be repeated to produce time intervals of any number or length. The output of each monostable can directly drive loads up to 200 mA, but it is generally recommended that the designer use transistor power stages such as suggested in Chapter 3.

Block 213   Multi-timing Using the 555 IC

Figure 13.8 shows how many monostable blocks can be wired to program combined time sequences in robotics and mechatronics applications. A single pulse (received from a remote control, sensor, or other device) initiates the timing process, and many devices can be activated for varying time intervals. Depending on the timing branches, many sequences can start at the end of the timing interval of each block.

Block 214   Bistable Block

The block in Figure 13.9 is very important in projects involving mechatronics and robotics. The first pulse applied to the input turns on the relay. The second pulse turns it off. The reset switch turns off the relay to initiate circuit operation with the relay turned off. This circuit is a flip-flop using the 4013 IC, a dual D-type flip-flop.

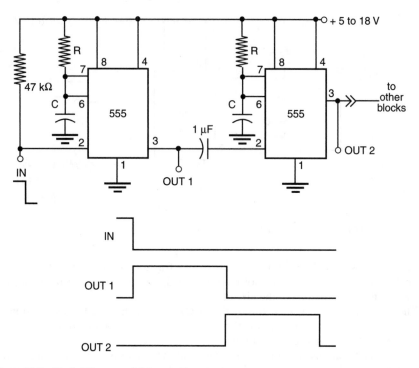

Figure 13.7   Block 212: sequential monostable.

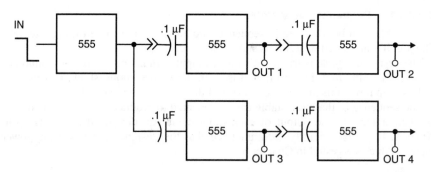

Figure 13.8   Block 213: multi-timing using the 555 IC.

This command signals can be produced by the on-off blocks described in previous parts of this book. In this block, the load is a relay that is driven by a transistor, but any of the blocks suggested in Chapter 3 can be connected to the output of the 4013. The reset switch can be replaced with a 0.1 µF capacitor if you want a power on reset (POR).

Block 215   Complete Bistable Using the 4013 IC

The pulses applied to the input of the previous block must be noise free and perfectly square. If your application operates with sensors or switches that can intro-

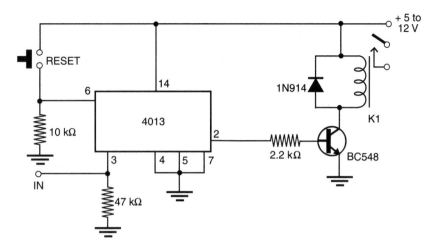

Figure 13.9   Block 214: bistable block.

duce noise into the circuit, a debouncer can be added. The complete block shown by Figure 13.10 includes a monostable to drive the 4013 with square pulses.

Each time S is pressed, the 555 produces only one pulse, changing the state of the circuit and turning the relay on or off. In the previous block, one of the flip-flops of the 4013 was used. In this block, we are using the other one (which demonstrates to the reader to observe the pinouts, since they are independent). This block can also drive other loads as suggested in Chapter 3. The time constant of the monostable can be varied to suit the application, to avoid bouncing, or for other purposes.

Block 216   Frequency Divider

In some of our projects, the designer may need a circuit that produces one pulse for each two or four pulses that are applied to the input. Flip-flops are ideal devices for

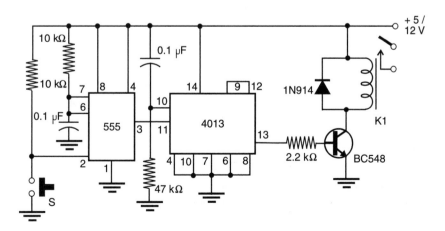

Figure 13.10   Block 215: complete bistable.

this task, and the 4013 is especially easy to use, as shown by the block in Figure 13.11. As with the previous block, this circuit can drive high-power loads directly if you use the blocks suggested in Chapter 3.

The input of this circuit must be a noiseless squarewave. If you are using sensors, the debouncer of Block 214 can be added. You can also wire several of these blocks in series to divide the frequency of a squarewave signal by any power of two.

Other ICs (including the 4020 and 4040) and configurations can be used to divide the frequency of a squarewave by other values. The ICs can divide the frequency of the signal by values up to 4k (4096) and 16k (16,384), respectively.

Block 217   Frequency Divider Using the 4020 IC

The 4020, a CMOS IC, can divide the frequency of digital signals by values from 2 to 16,384. This IC has 12 outputs at which the input signal is the input frequency divided by values of 2, 16, 32, 64, 128, 256, 512, 1024, 2048, 4096, 8192, and 16,384. The basic circuit using this IC is shown in Figure 13.12.

Notice that the circuit doesn't have outputs for division by 4 and 8. The outputs of this circuit can drive many of the blocks suggested in Chapter 3, and even a relay as shown in Blocks 213 and 214. It is important to remember that the duty cycle of the output signals is not 50%.

Block 218   Divide-by-10 Counter Using the 4017 IC

The block shown in Figure 13.13 is among the most important of those that use CMOS ICs. This circuit can be used in many configurations involving control and

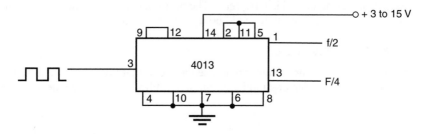

Figure 13.11   Block 216: frequency divider using the 4013.

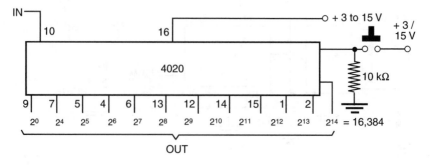

Figure 13.12   Block 217: frequency divider using the 4020.

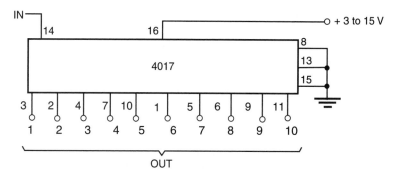

Figure 13.13   Block 218: divide-by-10 counter.

decoding. When the power is on, only one of the outputs goes to the high logic level, and the others remain at the low logic level. Each time a signal is applied to the input, the output in the high logic level goes back to the low level, and the next output passes from the low to high. All others remain in the low state.

When the last output is high and a new pulse arrives, this output goes low, the first goes to the high state, and a new cycle begins. To initiate the process by forcing the first output into the high state, we press the reset switch. The maximum operating frequency of this Block is 7.5 MHz with a 10 V power supply.

An important application for this block is in a sequential control. Each output drives a stage or a group of stages to initiate the functions to be performed. In the following blocks, we will see how this function can be programmed with resistor networks, simple diodes, and diode matrices.

Block 219   Divide-by-n Using the 4017 IC

The previous block can select one output of ten from a pulse stream applied to the input. In some applications, numbers other number than ten are required, and this block solves the problem. The circuit shown in Figure 13.14 can produce one output signal from n, where n is any number between 2 and 10.

To divide by n, we just need to connect the n + 1 output to the reset pin. For instance, as shown in the figure, if we want to have one high output from a sequence of three input pulses, we have to connect the fourth output (pin 7) to the reset pin (15).

The maximum operation frequency for this circuit is 7.5 MHz when the power supply voltage is 10 V. The outputs can drive stages suggested in Chapter 3 and a relay stage as shown in Blocks 213 and 214.

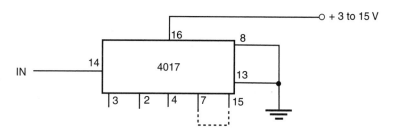

Figure 13.14   Block 219: divide-by-n.

Block 220   Programming with Diodes

Another way to program tasks for a robot, mechatronic arm, or other device is through the use of diodes as shown in Figure 13.15. The diodes are connected to source loads for different time intervals according to the nature of each load. Therefore, in the example, where we take the length of the input pulse as the time unit, OUT1 is high during the interval corresponding to two pulses, OUT2 is high during three pulses, and OUT3 during one pulse. You can also leave some outputs of the 4017 free if, in a particular instant, you do not want any load to be activated or any task performed.

The outputs can drive many of the blocks described in Chapter 3, and you can use two to ten outputs of the 4017. It is also possible to cacade the 4017 to have more than ten outputs.

Any silicon diode can be used in this circuit, and the power supply is in the range of 5 to 15 V. Although the 4017 operates with voltages as low as 3 V, we must consider the voltage drop of about 0.6 in the diodes.

Block 221   Digital-to-Analog Converter (DAC)

The circuit shown in Figure 13.16 can deliver an output voltage that depends on the number of pulses applied to the input. Starting with output 1 activated (pin 3), the voltage in the output is about 2/5 of the power supply voltage (consider the load to be a resistor and all others in parallel, since the corresponding outputs are at zero logic level).

When the second output is high (pin 2), the voltage falls to about 1/3 of the power supply voltage. When the last output is high (in this case, pin 1), the voltage is less than 1/50 of the power supply voltage.

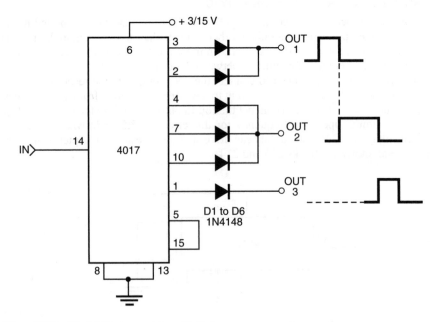

Figure 13.15   Block 220: programming with diodes.

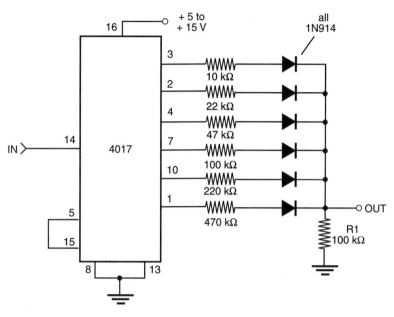

Figure 13.16   Block 221: digital-to-analog converter.

The circuit can be used to control a the voltage applied to a motor or a VCO in a PWM control. Since the output cannot deliver high currents, amplifier stages can be added. An operational amplifier is recommended to step down the output impedance and drive high-power loads.

Block 222   Sequential Programming Using a Diode Matrix

A sequence of operations can be programmed using a diode matrix as suggested by the circuit in Figure 13.17. The diodes determine which output is activated when the corresponding output of the 4017 is at the high logic level. For instance, when the first output (pin 3) is high (4017 reset), diodes D1 and D2 are forward biased, and positive voltages appear at outputs 1 and 3.

A sequence of operations for devices connected to outputs 1, 2, and 3 can be programmed by the position of the diodes in the lines. Using the 4017, it is possible to program from two to ten steps. The outputs can be expanded if more vertical lines or output lines are included.

An interesting idea to explore in our projects is to put the diodes on cards that can be connected to a slot in the circuit. You can then replace the card to obtain a different sequence of operations for the robot or mechatronic device. For an automatic elevator or arm, you can use different cards according to the sequence of stages or the desired operations. The command pulses can be generated by a single oscillator (a 555 astable, for instance) or produced by sensors or a keyboard.

Block 223   Voltage-Controlled Oscillator

Voltage-controlled oscillators (VCOs) can be used in many robots and mechatronic devices. For instance, you can use a VCO to produce pulses to control a pulse width

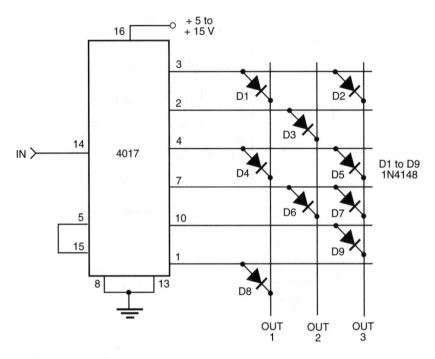

Figure 13.17   Block 222: sequential programming.

modulated (PWM) circuit with a linear sensor. You can also use a VCO to convert voltages into control signals for an electronic potentiometer.

The block shown in Figure 13.18 is a simple VCO using a CMOS IC. It can be used in many projects with other blocks presented in this book. This circuit produces a squarewave signal with a frequency that depends on the input voltage and also on the RC network. Typically, R can assume values between 10 kΩ and 1 MΩ, and C can assume values between 50 pF and 10 nF. For the values suggested in the diagram, the central frequency of the circuit is about 5 kHz. The output of this block can be used to drive many blocks suggested in previous chapters, including the PWM, linear controls, and others.

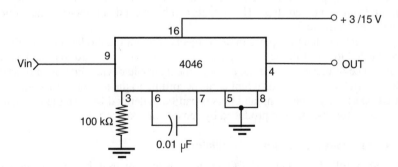

Figure 13.18   Block 223: VCO.

Block 224 Logic Switch

An important block for the mechatronics and robotics designer is a digital/analog CMOS switch that uses the 4066 (or 4016) IC. This device contains four analog/digital CMOS switches that can be used to switch a circuit on and off from a CMOS logic source. The switch is on when a positive voltage is applied to the control pin.

This circuit can operate in an analog or digital mode. In the analog mode, the IC is powered from a dual supply (+5/–5V), and a signal with amplitude between these limits can be controlled by the switch. In the digital mode, the circuit is powered from a single supply (+3 to +15 V and ground), and only digital signals can be controlled.

As shown in Figure 13.19, each of the four cells of a 4066 is formed by a switch that has a very high resistance when off (many megohms) and a low resistance when on (about 150 Ω). A pulse can be used to turn on the switch and drive a relay (as shown by the figure) or to control other blocks. The figure shows the pins of the IC for other switches.

Block 225 Diode Matrix

Diode matrices are the simplest forms of memory for programming a robot or mechatronic device. Figure 13.20 shows a basic block for a 4 × 3 diode matrix that can be used to program three outputs from signals applied to four inputs.

When a positive voltage is applied to input 1, diodes D1 and D2 are forward biased, and outputs 1 and 3 are activated. You can activate the inputs using sensors or other circuits used as a sequencer (Blocks 217 and 218). If a sequencer is used, it is possible, for example, to make a robot perform sequential tasks automatically.

The advantages in using of this type of memory are that (1) the diodes can handle high currents, directly activating loads, and (2) the matrix can be easily replaced by another, thereby changing the sequence or program.

Block 226 R/2R Network—DAC

The R/2R provides the simplest way to convert digital signals to analog. The simple block given in Figure 13.21 can convert the digital pulses produced by sensors, or by an IC such as the 7490, into corresponding voltages in a linear stair as shown in the figure. Since this is a high-impedance circuit, you'll need some kind of amplifier. In Block 227, we give the complete DAC with this amplifier.

The values of the resistors depend on the input voltage and the output impedance. Typically, R can be between 1 k and 10 kΩ. A DAC such as this can be used to pro-

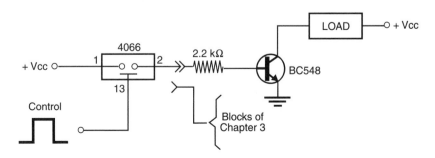

Figure 13.19 Block 224: logic switch.

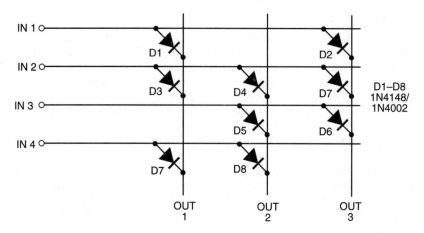

Figure 13.20   Block 225: diode matrix.

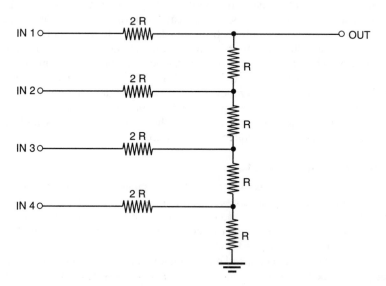

Figure 13.21   Block 226: R/2R DAC.

vide analog control of a load (using one of the blocks described in the previous part) from digital blocks.

Block 227   Binary Coded Decimal Counter/Decoder

The 7490 (TTL) is another important IC that can be used in many projects involving pulse counting. The basic block for a binary coded decimal (BCD) is shown in Figure 13.22.

The circuit must be powered from a 5 V supply, and the outputs are activated as indicated in the following table.

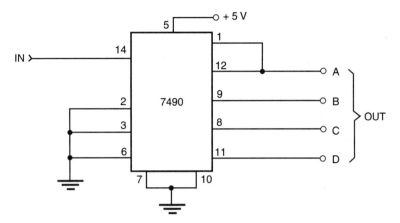

Figure 13.22   Block 227: BCD counter/decoder.

Input Pulse	D	C	B	A
0	0	0	0	0
1	0	0	0	1
2	0	0	1	0
3	0	0	1	1
4	0	1	0	0
5	0	1	0	1
6	0	1	1	0
7	0	1	1	1
8	1	0	0	0
9	1	0	0	1

This block can be used to drive Block 225, converting input pulses into voltages.

Block 228   Complete DAC

A complete DAC using an R/2R network and an operational amplifier is shown in Figure 13.23. This circuit can be used to convert BCD information into voltages in a range from −1.4 to +1.4 V. This value is determined by R1. The value can be increased if you want another range of output values, but remember that the range cannot be greater than the power supply voltages.

The indicated IC is the 741, but equivalents can be used. Observe that the circuit needs a dual power supply. The output impedance for this circuit is about 150 Ω. This means that it can drive many of the power blocks described in the initial parts of this book.

Block 229   64-Bit RAM

The block shown in Figure 13.24 is a 16-word × 4-bit random access memory (RAM) that uses a single TTL integrated circuit. The robotics and mechatronics designer can program a sequence of 16 words of 4 bits each to command a device and

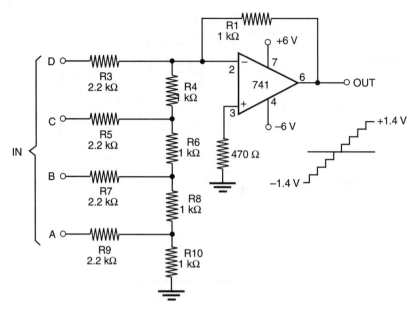

Figure 13.23   Block 228: complete DAC.

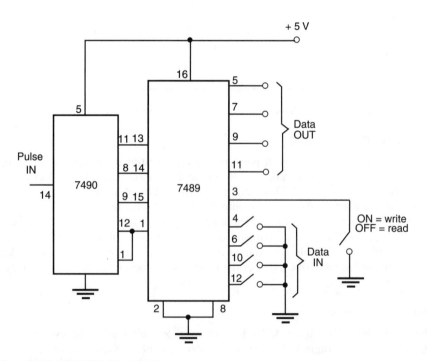

Figure 13.24   Block 229: 64-bit RAM.

enable any of these commands simply by addressing the position of the memory where it is stored.

The data out pins (5, 7, 9, 11) can be used to control any of the blocks suggested in the previous chapters, including the ones using relays or directly driving high-power loads. To store the data, use the following procedure:

- Initialize the address inputs to the position in memory in which you want to store the data (pins 13, 14, 15, and 1).
- Send the data to the data-in pins (4, 6, 10, and 12).
- Turn switch S on and off.
- Input the address for the next data.
- Send the next data to the data-in switches.
- Close and open switch S again.

Reading from the memory is very easy. You just need to access the data via the address inputs and leave S open. The stored data will appear in the data-out lines.

13.5   Additional Information

Some integrated circuits used in the blocks of this part are very important for the robotics and mechatronic designer. Knowledge of their characteristics is fundamental for creating new projects. Therefore, in this section we give some additional information about three particularly useful ICs.

4017

The 4017 is a decade counter with 10 outputs. It can be programmed to count pulses between 2 and 10 (as shown in the blocks). Each output of the 4017 can drain or source 2.25 mA when the circuit is powered from a 10 V power supply. The maximum operating frequency is 5 MHz (typical). The pinout for this IC is shown in Figure 13.25.

4046

The 4046 CMOS IC is a phase-locked loop (PLL) that basically can be used for the same tasks as the LM/NE567 (see previous chapters for more information). Figure 13.26 shows the pinout of this IC.

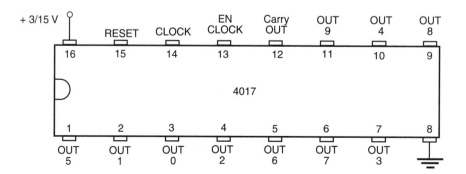

Figure 13.25   The 4017 IC.

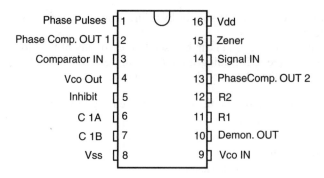

Figure 13.26   The 4046 PLL.

7490

The 7490 (Figure 13.27) is a TTL decade counter (divide by 10). This circuit is formed by two stages (a divide-by-two and a divide-by-five), and the outputs are binary coded decimal (BCD). The count advances on the negative changing input signal.

13.6   Suggested Projects

Many interesting projects can be created by combining blocks from this chapter with blocks from other chapters. Some of the combined projects can be complex, since the use of logic can add a degree of "intelligence" or automation.

In some cases, the following suggestions require minor changes in the blocks' components to match their characteristics and achieve maximum performance.

1. Figure 13.28 shows a combination of blocks used to create an automation circuit that is applicable to robots and mechatronics devices. Three or more sensors send the signals to an OR block (Block 207). The signal of any sensor is sent to a monostable block (Block 210) that produces a short pulse to the next block (Block 221). This circuit counts the pulses and activates two other blocks. Block 52 controls the direction of a motor, and Block 46 inverts the direction of the other motor and stops after a predetermined interval.

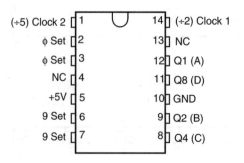

Figure 13.27   The 7490 IC.

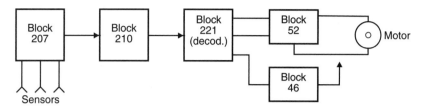

Figure 13.28   A multi-block project.

2. Figure 13.29 shows a design that uses many of the blocks shown in this book. It is a remote controlled system that is applicable to robots and other projects.

Block 193 codes tones for three channels, and these tones are applied to a radio transmitter (Block 203). The signals are picked up by a receiver (Block 205) and sent to a tone decoder (Block 194), resulting in three channels of control.

One of the channels is used to control a bistable (Block 213) that reverses the direction of a motor. The other channel is used to control a sequence of events using programmed Blocks 219, 220, or 221, and the third channel is a direct actuator controlling a load (Blocks 30, 31, or any other) for this task.

13.7   Review Questions

1. What is the difference between a monostable and a bistable multivibrator?
2. How can logic decisions be made using logic gates?
3. How can a counter be used to perform logic tasks?
4. How can we program multiple time intervals using the monostable 555?
5. What is the 4013 CMOS IC?

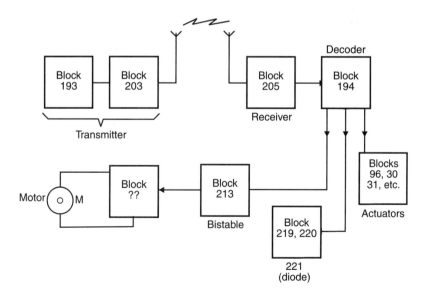

Figure 13.29   Another project using many blocks.

6. What is a DAC, and where it is used?
7. What is a VCO, and where it is used?

14

Intelligence and the Computer

14.1 Purpose

The purpose of this chapter is describe how can circuits and programs be used to add intelligence to robots and mechatronic devices. It presents an introductory concept of intelligence as applied to mechatronics and robotics and shows how it can be added using software or hardware. The reader will also see how circuits and programs can "learn." The parallel between the intelligence of living beings and machines will be also discussed.

14.2 Theory

14.2.1 What is Intelligence?

This is not an easy question to answer, considering that researchers are not even sure how our own brain operates. However, natural intelligence normally can be associated with the ability to

- Learn from experience
- Make logic decisions based on experience
- Generate emotions

Of course, it is not easy to install these characteristics in an electronic system (especially the third one), but the basic concepts at least can be mimicked using circuits and programs. Circuits and programs that embody the above characteristics are normally associated with *artificial intelligence.*

And, of course, the degree of intelligence that a system possesses constitutes another problem. It is not easy to measure the intelligence of a machine. Evaluating whether a robot has even the intelligence of a worm is a complex undertaking. However, if a circuit can produce something through the process of making its own decisions, then it is valid to classify it as "intelligent," and this concept is explored in this chapter. The basic idea is to create blocks that can react and learn. We then progress to simple circuits that can put this intelligence to practical use. Because complex circuits are always derived from simple ones, we are creating a base from which much more interesting projects can be developed.

14.2.2 Intelligence by Hardware and Software

There are two ways to add intelligence to a robot or a mechatronic project. The first possibility is to use a computer or microprocessor that is programmed to provide artificial intelligence characteristics to the device. In this case, the intelligence is in a program that consists of commands and instructions that can modify themselves (learn) as a reaction to sensory input. We call such systems *intelligent by software.* Special programs are constantly being developed to allow systems to learn. Many artificial intelligence programs run on common microcomputers and can be used by amateurs and students.

The second way to add intelligence to a robot or mechatronic device is to employ circuits that learn and change their characteristics through experience. Such circuits are said to employ *intelligence by hardware.* The *electronic neuron* is the usual type of circuit used to add intelligence by hardware.

14.2.3 Electronic Neurons and Neural Networks

The basic building block of natural intelligence is the neuron. The neuron is a cell, illustrated in Figure 14.1, that has inputs (dendrites) and an output. When excited by stimuli (temperature, light, sound, etc.), the neuron produces a train of electric pulses (about 50 mV) at its output.

In the nervous system, including the brain of complex creatures such as humans, billions of neurons are interconnected, and from this complexity comes intelligence. The ability the nervous system, including the brain, to learn is based on the fact that the performance (output) of neuron changes according to the number and intensity of pulses it receives. In other words, the neuron can be conditioned by (learn from) an experience. If a particular type of input tends to arrive with a specific power and frequency, the neuron adapts to it and tends to produce outputs that correspond only to that type of input.

A simple neuron is not difficult to implement in electronics. Many simple configurations can imitate the performance or a living neuron, and when these neurons are interconnected in a *neural network,* they can add artificial intelligence to a project.

Today, neural networks are the subject of intensive research. Scientists believe that eventually they will be adapted to robots and other electronic devices to give them something that resembles natural intelligence, including the ability to learn, have emotions, and experience a human-like "way of life." Accomplishing this, however, will require neural networks that contain billions of circuits or blocks, which is not possible with contemporary technology. Today, we will have to be satis-

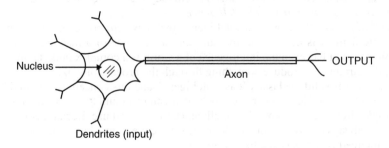

Figure 14.1 The neuron.

fied simply to add some electronic neurons to some of our projects. This will allow them to learn simple tasks, but with a degree of intelligence that still does not measure up to that of a worm.

14.2.4 Fuzzy Logic and Intelligent Software

The traditional digital logic that consists of a "yes" or "no" decision is not good enough for projects involving intelligence. If we adopt a third condition that corresponds to an answer of "perhaps," we have a better approach to something that can be called intelligence. This is the basis of *fuzzy logic.*

The idea of the fuzzy logic is just as the name implies. To give a device more than the traditional two choices, we have to design a circuit that can transition to more states than the two states (1 and 0) of digital electronics.

Fuzzy logic can be employed in special programs that are suitable for artificial intelligence. Many books are devoted entirely to the subject, and the reader may want to consult them at a later date, as they will make it possible to add even more sophisticated levels of intelligence to the blocks shown in this chapter. Adding fuzzy logic to these blocks can result in intelligent machines.

14.2.5 Microcontrollers and Microprocessors

Robotics and mechatronics designers can find many microcontrollers, microprocessors, and even digital signal processors (DSPs) that are suitable for these projects. Using this kind of circuit requires programming skills and a thorough knowledge of the architecture of each type of device. Therefore, we recommend that readers who need to know more about them seek out courses or books that specifically address the subject.

PIC[*] devices, and the BASIC Stamp[†] in particular, have characteristics that are ideal for projects in this area. The Basic Stamp is a small microprocessor that can be programed in BASIC using any PC to perform simple tasks and to read information for sensors. Figure 14.2 shows a BASIC Stamp.

The small printed circuit board includes a space where the designer can implement interface or sensor circuits. This means that you don't need additional boards to build a simple project involving a BASIC Stamp, because it can accommodate the other circuits on its own board.

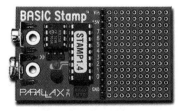

Figure 14.2   BASIC Stamp®.

14.2.6 Personal Computers

The computer can be used as a powerful tool in mechatronics and robotics projects. The computer can communicate with external devices using its serial and parallel ports. Through these ports, the computer can control external devices or receive information from sensors or other circuits. This information can be used for control functions or employed by programs that add intelligence to the device.

Using the PC in these projects is not all that easy. The designer needs to know what kind of devices can be plugged into the PC and what kind of signals it can send and receive. To connect a computer to a device, the designer needs three things:

1. An interface that converts the signals produced by the PC into a signal that can be used by the device.
2. An interface that converts the signals produced by the device into a signal that the PC can understand.
3. Software to produce or interpret the signals.

There are many simple ways to access the input/output (I/O) ports of a PC. By writing a few lines in BASIC, VisualBASIC, Delphi, C, or another programming language, it is possible to read the logic levels that exist in the serial or parallel port or to send signals to these ports.

The main consideration when connecting any device to a PC I/O port is possible damage to the equipment. If something goes wrong with the external device, there is a high likelihood that it will cause damage to computer components. If something is burned out in the PC, the repair cost may be high.

To avoid this danger, it is necessary to use isolated interfaces. Blocks of this kind will be presented.

The blocks included in this chapter will all you to interface your computer with many of the other blocks described in this book, enabling computer control of many robotic and mechatronic devices.

14.3   Using the Hardware

Many electronic circuits can be used to interface a mechatronics or robotics project with a computer or to add intelligence. Many blocks are simple and require only a few intelligent components, but others are complex, requiring a computer. In the next sections, we will show the reader some simple blocks that can add decision-making capabilities, memory, and (in some cases) traces of we can call *artificial intelligence.*

14.4   Blocks

The following blocks are not intelligent. However, depending on the way they are used, they can add intelligence to our projects. The implementation depends on the designer's preferences, and suggestions will be given in each case. The designer is free to make changes or adapt any block to the specific task of interest.

Block 230   Learning Circuit Using Power MOSFET

The simple circuit shown in Figure 14.3 changes its output according to "experience" (in this case, the number of pulses applied to the inputs). The pulses are produced by momentary contact switches (relays or sensors) connected to IN1 and IN2.

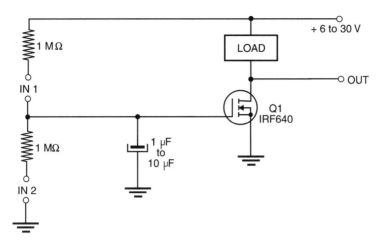

Figure 14.3   Block 230: learning circuit using power MOSFET.

Any short pulse applied to IN1 pumps a small charge into capacitor C, increasing the output voltage. The pulses applied to IN2 drain some charge from the capacitor, decreasing the output voltage.

The output voltage can be used as reference in a block comparator or other circuit that will change its performance according to the number of pulses that have been applied to the inputs. We can say that, depending on the pulses applied to the inputs, the circuit can be *conditioned* or can *learn* and generate an appropriate response.

The number of pulses required to teach this circuit depends on the value of the capacitor. This component can assume values in the range between 1 and 10 μF. Polyester capacitors are recommended, because their charge retention time is longer than that of electrolytic capacitors. It is important to remember that this circuit is not linear, and the output voltage jumps to near zero when the power FET reaches the saturation point.

The circuit can directly drive high-power loads. Replacing the load with an 1 kΩ resistor, the output voltage can be used as a reference for comparators.

Block 231   Learning Circuit Using an IC

The great advantage of the block shown in Figure 14.4 is that it is linear. The very high input impedance of the JFET operational amplifier (CA3140 or equivalent) adds long-term memory to the circuit. The output voltage, on the other hand, will be the same as the voltage found between the capacitor's terminals.

In this version, the pulses applied to the input pump charges into the capacitor, increasing the voltage applied to the operational amplifier (opamp). The opamp is wired as a voltage follower, which means that the output voltage is the same as applied to the input. With a few alterations, this circuit can operate with two inputs, as in the case of Block 230, learning when signals are applied to IN1 and forgetting when the signals are applied to IN2.

Block 232   Sample-and-Hold Circuit

Sample-and-hold circuits can be used as intelligence blocks in many projects. The circuit shown in Figure 14.5 operates as follows:

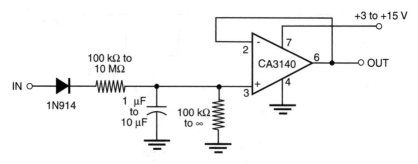

Figure 14.4   Block 231: learning circuit using an IC.

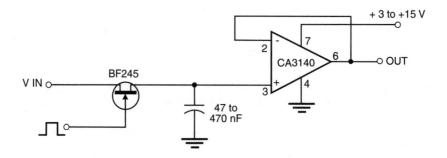

Figure 14.5   Block 232: sample-and-hold circuit.

- If a control voltage is applied momentarily to the gate of the transistor, it turns on, and the capacitor is charged with the voltage that then exists at the input.
- After the transistor turns off, the capacitor retains the information about that voltage, using it to drive the opamp, which produces that voltage at its output.
- If the control signal is a sequence of pulses, the voltage that exists at the output of the circuit will change according to the voltages in the input at the time each pulse is produced.

One application of this circuit is to use it as memory to store information about the voltage in a sensor at any particular moment.

Block 233   High-Power Learning Circuit

The blocks controlling a CA3140, as we have seen, can control high-power loads (up to 3 A) if the configuration shown in Figure 14.6 is used. The reference voltage in the output of the CA3140 is applied to the LM350, resulting in a voltage output that is 1.25 V higher. This block can control loads up to 3 A, but the LM350 must be mounted on a heat sink.

Block 234   Light-Conditioned Circuit

The block shown in Figure 14.7 can be used in experiments involving circuits that learn. The reference voltage depends on how much light falls on the LDR. In a

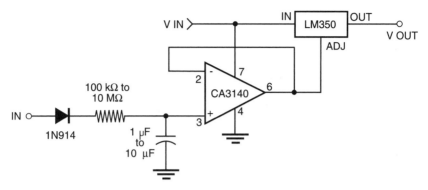

Figure 14.6   Block 233: high-power learning circuit.

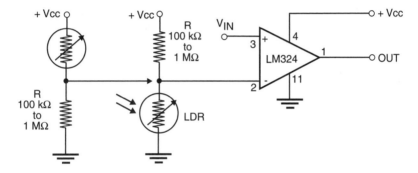

Figure 14.7   Block 234: light-conditioned circuit.

project involving artificial intelligence, this block can be used to change the response of the circuit according to light levels in a particular location.

If the LDR is located near a small lamp (forming a optocoupler), the lamp can be driven by a control circuit, thereby determining the voltage reference. If the voltage in the lamp changes, the reference voltage changes, too. Other operational amplifiers and comparator can replace the LM324 in this block.

Block 235   Thermal Memory

This is a very interesting block for experiments involving conditioned circuits. It is illustrated in Figure 14.8.

The wirewound resistor is placed inside a thermal can with a negative temperature coefficient (NTC) device. Each time a pulse is applied to the resistor (coming from a control circuit or a sensor), the small amount of heat that is produced raises the temperature of the water in the can. Since the resistance of the NTC, and therefore the reference voltage, depends on the temperature and the number of the pulses, this circuit can be "conditioned" by pulses applied to its input.

For experimental purposes, a 22 $\Omega \times 1$ W resistor (R2) receiving pulses from a 12 V source (about 500 mA) can be used with about 50 mL of water in a thermal can. The value of resistor R1 depends on the desired reference voltage at ambient temperature. For experimental purposes, it can have the same value as the NTC.

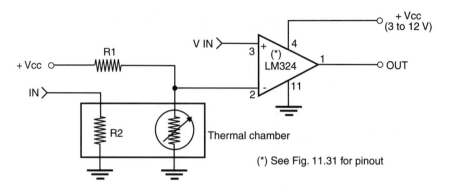

Figure 14.8   Block 235: thermal memory.

The "memory" of this circuit depends on how fast the thermal can dissipates its heat. If a good thermal material is used, the circuit can "remember" the amount of applied pulses for several hours. Of course, a robot made with this type of block will have to relearn its commands and responses from one day to the next.

Block 236   Teachable Window Comparator

The previously shown blocks using thermal cells (Block 235) or capacitor cells (Blocks 230 through 232) can be used in experiments involving electronic neurons. These devices can change the bandwidth of a stimulus passing across it according to a training process. Figure 14.9 shows a thermal memory neuron using a window comparator.

Three cells are used to change the voltage bands of the window comparator. Depending on the pulses applied to the resistor inside the cells, the temperature will change, altering the reference voltages of the window comparator.

The pulses are produced by sensors or other control circuits. For example, if pulses are produced that increase the temperature only of R2, the bandwidth of volt-

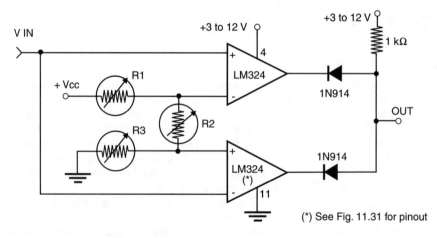

Figure 14.9   Block 236: teachable window comparator.

ages recognized by the circuit will be narrowed, as shown in Figure 14.10. Pulses applied to R1 will reduced the upper bandwidth of reference voltages.

Configurations using one or two thermal memory cells can be also used in this block. The same circuit can be used with other conditioning elements such as LDRs.

Block 237   Electronic Neuron (I)

The circuit shown in Figure 14.11 is an electronic neuron that can learn, changing its performance according to the number of pulses applied to the input. A pulse at the input simultaneously adds a small charge to C, altering the reference voltage. If the charge becomes powerful enough to reach the trigger threshold, it triggers the LM324, charging the capacitor at its output. The discharge of this capacitor across the 555 produces a train of pulses at the output. This performance imitates a real neuron, which also produces pulse trains when stimulated.

The values of the components shown in the circuit are average. The reader must make changes to achieve the desired performance.

Block 238   Electronic Neuron (II)

Another electronic neuron, shown in Figure 14.12, uses the capacitor charge as memory and two 555 ICs to produce a pulse train when stimulated by one or more pulses. The train of pulses at the output of this circuit also depends on the time constant of the RC circuits and the threshold of operation of the charge in capacitor C.

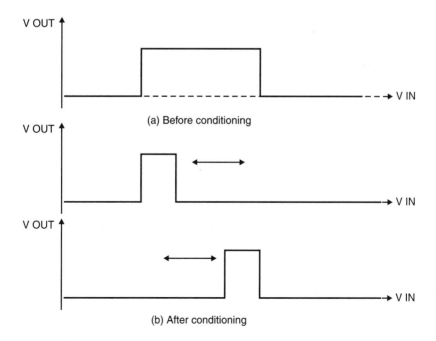

(a) Before conditioning

(b) After conditioning

Figure 14.10   Changing the response.

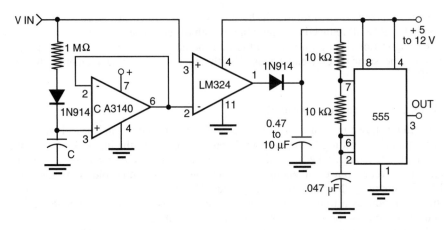

Figure 14.11   Block 237: electronic neuron I.

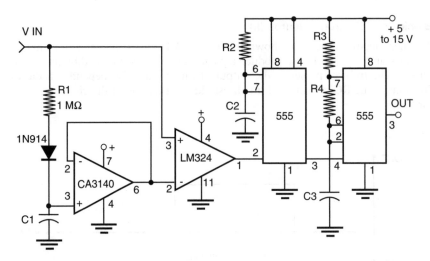

Figure 14.12   Block 238: electronic neuron II.

Block 239   Learning Circuit Using the 4017 IC

Figure 14.13 shows an interesting circuit that can be taught to recognize ten different input voltages. If the voltages are generated by sensors, the circuit can be conditioned as a neuron.

Starting from the point at which the output voltage is low, the circuit can recognize pulses of almost all amplitudes. But, at each pulse, the sensitivity of the circuit decreases, since the 4017 changes its output, altering the reference voltage for the comparator.

Block 240   Integrate-and-Fire Neuron Using Operational Amplifiers

The circuit shown in Figure 14.14 is the traditional integrate-and-fire version of an electronic neuron using operational amplifiers. The circuit is formed by an integrator

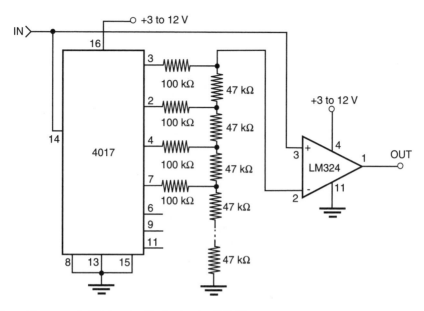

Figure 14.13 Block 239: learning block using the 4017 IC.

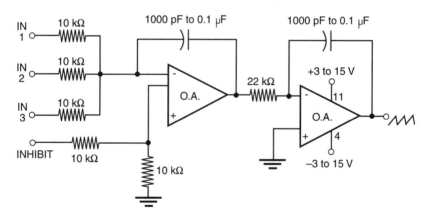

Figure 14.14 Block 240: integrate-and-fire neuron.

and a comparator and has an inhibit input. The components have values for operation in the 500 Hz range. Note that this circuit operates with a pulse train.

14.4.1 Connecting a Computer

The following blocks are interface circuits that allow the reader to connect a computer to robots and mechatronic devices. The blocks are basically designed to use the parallel port. A reader who wants to build more sophisticated projects, using the serial port or a greater number of communication channels, should refer to specific literature dealing with that subject.

Block 241   Simplest Parallel Interface

Figure 14.15 shows one of the simplest ways to send signals from the PC to a relay that controls an external load. Using this circuit the designer can control up to eight relays by addressing the outputs of the parallel port. When the output is set to the high logic level, the relay turns on and controls an external circuit.

Two important points must be considered with regard to this circuit:

- The circuit is not electrically isolated from the PC, so if something goes wrong with the block, the PC can be damaged. Do not use this block if you don't fully understand the possible consequences!
- The PC and power supply of the electrical stages use a common ground.

Block 242   Parallel Interface (II)

The interface shown in Figure 14.16 is very sensitive and doesn't load the outputs of the parallel port. The circuit can be adjusted for the best sensitivity using each trimmer potentiometer, and you can use only two LM324 devices to drive eight loads, since each IC has four comparators.

A variety of blocks can be driven from the output, as suggested in the diagram. If the circuit is powered from a 5 V supply, the circuit is compatible with TTL logic.

Block 243   Isolated Interface (I)

The main problem with all of the interface blocks described previously is that they are not isolated from the PC. If something goes wrong with the circuit, the PC can be damaged. This block is ideal for the designer who wants to protect the computer from faults and shorts. The block shown in Figure 14.17 is completely isolated from the PC.

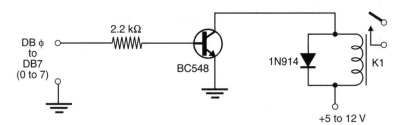

Figure 14.15   Block 241: simplest parallel interface.

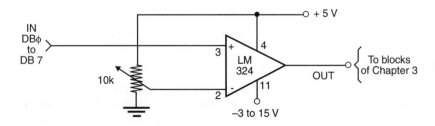

Figure 14.16   Block 242: parallel interface II.

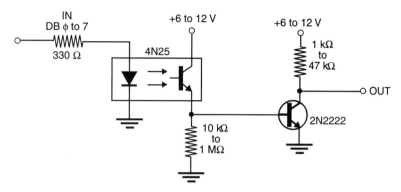

Figure 14.17   Block 243: isolated interface I.

The optocoupler is the 4N25, but any equivalent can be used. The reader must experiment to obtain the best values for the resistor as appropriate to the blocks to be driven and the power supply voltage. A range of values for the resistors is indicated in the diagram.

Block 244   Isolated Interface (II)

The circuit shown in Figure 14.18 is the best of all interfaces, since it combines the isolation of an optocoupler and the sensitivity of an integrated comparator. In the basic circuit, P1 can be adjusted to obtain the best sensitivity, but in normal conditions you can replace this component with a fixed-resistor voltage divider. The reference voltage will be about 1/3 of the power supply voltage, meaning that a 22 kΩ and a 10 kΩ resistor will operate well in most practical applications.

When powered from 5 V supplies, the circuit is TTL compatible. For other voltages, many of the blocks described in Chapter 3 can be driven.

Block 245   Interface for AC-Powered Loads

The MOC3010 is an optocoupler with an optodiac inside. This means that this device can be used to directly control a triac with a load powered from the ac power line.

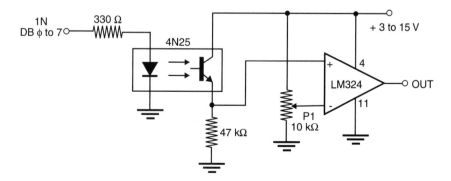

Figure 14.18   Block 244: isolated interface II.

The block shown in Figure 14.19 is a typical application for this device. Applying a high logic level to the output of the parallel port, the optodiac turns on the triac, powering a load connected to the ac power line. If you intend to drive loads from 220 Vac or 240 Vac power lines, the optocoupler or optodiac to use is the MOC3020.

Using a TIC226, loads up to 6 A can be controlled from the PC. Since the optodiac has a typical isolation voltage of 7,000 V, you don't need to worry about faults in the controlled circuit affecting the PC.

Block 246   Data Acquisition Interface Using Comparators

Voltage comparators such as found in an LM324 can be used to transfer data from a sensor to a PC via the parallel port. The typical block showing one channel of data acquisition is illustrated in Figure 14.20. It is important to remember that this block must be powered from a 5 V power supply and use common ground with the PC.

All of the previously described blocks using resistive sensors and on-off sensors with the LM324 can be used to send information to the PC via the parallel port. In the figure, the application shows how the block can be used in an "electronic eye" for a robot that uses an LDR as a sensor.

Block 247   Data Acquisition Interface Using the ADC0808

The ADC0808 is a single chip analog-to-digital converter with outputs that are compatible with the characteristics of the parallel port of a PC. This IC has eight analog

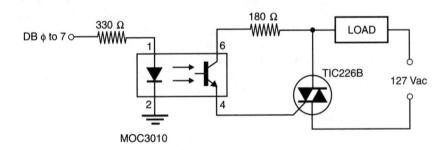

Figure 14.19   Block 245: interface for ac-powered loads.

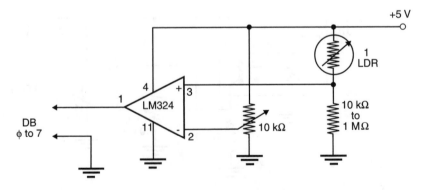

Figure 14.20   Block 246: data acquisition interface with comparators.

inputs where voltages between 0 and 5 V can be converted into digital form, giving values between 00000000 and 11111111 (0 to 256 steps).

As shown by the block in Figure 14.21, it is necessary to apply a 500 kHz digital signal (clock) to the conversion input. (An oscillator using a 4093 IC can be used for this task.)

In the input select, we apply the logic signals that determine which analog input is active at a particular time. The circuit must be powered from a 5 V power supply.

14.5 Additional Information

Because there is not enough space in this book to provide complete information about all possible projects in this area, it is important for the reader to seek out other available literature. Various sources provide guidance about more complex blocks, the use of microprocessors, and applications for the BASIC Stamp.

The BASIC Stamp® is manufactured by the Parallax, Inc., and this microcontroller is very easy to use in robotics and mechatronics projects. The great advantage of this small microprocessor is that it is complete; it can run BASIC programs, and it has built-in I/O ports, timers, A/D converters, and serial communications. Much more informations about the BASIC Stamp can be found on the company's web site, which can be accessed at http://www.parallaxinc.com.

PIC® processors, from Microchip Technology, Inc., are very popular among builders of robots and mechatronics devices. They can be programmed by a PC and

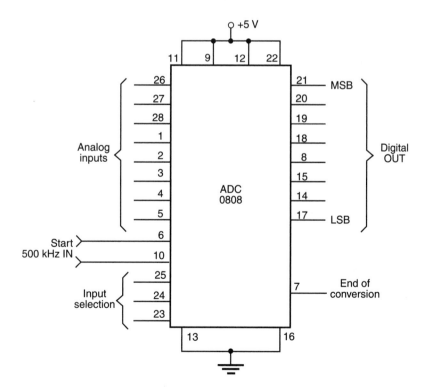

Figure 14.21 Block 247: data acquisition interface using the ADC0808.

erased by a UV source, and new versions have flash memories that can be programmed and erased electrically. They are available in different sizes and with a range of options. The URL http://www.microchip.com will take you to the company's web site.

14.6   Other Microcontrollers

Other families of microcontrollers can be used in robotics and mechatronics. One of them is the 8051 family. Many manufacturers include 8051 products in their lines, since they are used in industrial applications such as in the automotive industry.

14.7   Suggested Projects

Using the blocks described in this part and the others seen in the previous chapter, our projects can rise to a sophisticated level. The reader can build intelligent robots, computer-controlled robots (via wires or radio), and projects that can learn.

The following suggestions are among the simplest and, of course, the reader is free to make changes or to add other blocks and functions.

1. Build a robot that follows a light source but also can become "tired" if the light is too powerful. Design this robot to walk faster if the light intensity is reduced.
2. Make an experimental circuit that controls a simple relay or lamp and that can learn from the number of pulses applied by switches. (Use the thermal window comparator.)
3. Design an elevator that can be controlled by a PC.
4. Using LDRs, create an electronic eye that can send the image information to a PC.
5. Use the sound output of the PC to control a device. Generate tones using software (different tones emitted when different keys are pressed) and decode the signals using the Blocks 192 and 194. The tones can be sent from the PC to the device using the transmitter of Block 203 and received by Block 204 or 205.

14.8   Review Questions

1. What is intelligence?
2. What is the difference between intelligence by hardware and by software?
3. How can a circuit learn?
4. What is a thermal memory cell?
5. How can an electronic neuron learn?
6. How can a microprocessor be used to add intelligence to a robot?
7. What is a neural network?
8. What is an ADC?
9. How many outputs does a parallel port have?

15

Other Blocks: Light and Sound Effects

15.1 Purpose

The purpose of this part is provide some miscellaneous additional blocks that can be useful in our mechatronic and robotic devices. The blocks include sound effects, light effects, power supplies, and battery chargers. In this chapter, we will offer some suggestions about how these blocks can add more realism to the projects.

15.2 Theory

Many electronic circuits can be used to drive lamps, producing visual effects for robots and mechatronic devices. There are also many circuits that can produce sounds, and these are suitable for the same projects. Other important blocks add some defensive capabilities to a robot, allowing it to fend off attacks, and others are used to power the electronic circuits. All of these types will be studied in this chapter.

15.2.1 Light Effects

The main light effects that can be used in these projects are described below.

Flashers. Flashers are the most common light effects added to robots and mechatronics. These circuits cause LEDs or lamps to flash at a rate that can be fixed or variable by sensors or the circuits' operation. Simple circuits basically consist of a low-frequency stage driving a load.

Stroboscopic lamps. A low-frequency oscillator driving a lamp at frequencies in the range of 0.05 to 0.2 Hz produces a stroboscopic effect. The fast flash of a lamp creates the effect of freezing movement. A stroboscopic lamp in a robot can draw much attention to it.

Inverters. Inverters are circuits intended to produce high voltages from low-voltage dc power supplies. These blocks can be used to power fluorescent lamps from batteries or cells.

The basic inverter is a low-frequency oscillator driving a transformer. Small fluorescent lamps in the range from 4 to 10 W can be driven by simple circuits that use one or two transistors.

Sequencers. Sequencers are circuits that drive lamps or LEDs sequentially. The most popular of the ICs used for this task is the 4017, which was described in the previous chapter. A sequencer that drives three to ten LEDs or lamps can add interesting special effects to these projects.

Bar and dot indicators.   Bar indicators are circuits that form visual scales with LEDs or lamps. A number of lamps or LEDs, depending on the input voltage, are turned on to vary the bar's length. For example, if you apply 2 V, two LEDs glow. If 5 V is applied, five LEDs light up, and so on.

Dot indicators, on the other hand, use the same disposition of LEDs or lamps, but only one LED is on when the input voltage is applied. Figure 15.1 shows the two types of displays.

Dot and bar displays are easy to build using the LM3914 IC. This IC can operate in both modes, depending only on the external connection of pin 3. Block 254 will show how this IC is used.

15.2.2 Sound Effects

Sound effects include sirens, voice synthesis, machine gun sounds, etc. Simple oscillators can be added to generate common sounds such as sirens, but more complex blocks using dedicated ICs can perform sophisticated sound synthesis and provide other effects.

Since this book is intended for students and experimenters, we show only simple blocks that can be implemented with few cheap components. It is up to the reader move on and design more complex devices that include, for example, voice synthesis by the computer.

The transmitter and receiver blocks (Blocks 203 and 205 and small FM receivers) can be used for wireless transfer of computer voice synthesis to a robot or other mechatronic device.

Note: The reader can find many circuits that use the 4093 IC for light and sound effects described in the book, *CMOS Projects and Experiments* (Newnes).[*]

15.2.3 Other Blocks

Functions that can be included in a project depend only the range of our imagination. There are essentially no limits, and some unusual ideas can be created with blocks involving sounds, high voltage, light, and so on.

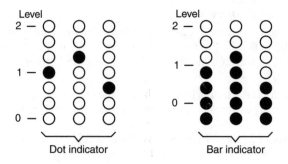

Figure 15.1   Dot and bar indicators.

---

* ISBN 075067170X; visit www.newnespress.com for information.

Some blocks that are not used directly in a robot or mechatronic device are nevertheless important in supporting its operation. These include power supplies and battery chargers.

15.2.4 Self-Defense

Of course, you have no intention of building an Armageddon machine, a device that projects an array of deadly laser beams, or a weapon that electrocutes an enemy with a one-megavolt discharge. However, for demonstration purposes or for some measure of protection, it is possible to add some defensive resources. There are many possibilities, but here we focus on only two:

1. *Electrical discharge.* It is possible to generate a high voltage from the device's internal batteries. Because we have a limited amount of current available, the discharge will not be dangerous, but simple circuits can generate shocks with voltages between 600 and 10,000 V.
2. *Powerful audible alarm.* A powerful alarm can be used to get the attention of anyone who touches a robot. Many simple circuits can drive small (but powerful) loudspeakers, producing modulated tones. An interesting adaptation is the use of ultrasonic sounds to scare away animals.

15.2.5 Power Supplies

If a robot or mechatronic device is to be powered from the ac power line, special circuits are needed to convert the ac line voltage to the dc needed by the circuits. These circuits are called *power supplies* and are required blocks for any ac-powered project.

The basic power supply must deliver a fixed voltage that is appropriate for the project (3 to 12 V, typically) and enough current to supply all the circuits in the device. In most cases, currents between 500 mA and 3 A are sufficient, but heavy-duty projects can require more current. The blocks shown in this chapter will provide enough power for most of the projects you will create.

15.2.6 Battery Chargers

Common cells are expensive and can't produce high currents over long periods of time. If the project is dc powered, drains high currents, and can't be powered from the ac line by a power supply, you should consider using rechargeable batteries.

There are many types of rechargeable batteries, including NiCad and gel cells. These cells can supply a great deal of energy to a robot and can be recharged many times. But to recharge such batteries and cells, you need a battery charger.

A battery charger is a simple power supply that forces current to flow in the reverse of its normal direction (see Figure 15.2). If you let the current flow for a while (6 to 16 hours for common types), the battery or cells will be recharged.

When charging batteries or cells, it is important to limit the current flow to the maximum recommended level. *Excess current can damage a battery or even cause it to explode!* Some of our blocks will show simple chargers for battery types that are suitable for our robotic and mechatronic applications.

15.3   Blocks

The following blocks can either be used alone, having some independent effect on a robot or mechatronic device, or controlled by other blocks that use relays, sensors,

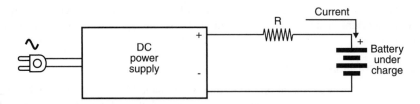

Figure 15.2   Battery charger.

or other control blocks. In some cases, the components must be varied to achieved the best performance. The reader is free to experiment to find the most favorable effects in each case, altering not only the component values but other parameters including supply voltage and even component type.

Block 248   Low-Voltage LED Flasher

You can add realism or special effects to robots and mechatronic devices by adding flashing LEDs. The block shown in Figure 15.3 is important when the robot or mechatronic project has no leftover space in which to install conventional flasher circuits.

This block is an LED flasher that can operate with voltage supplies as low as 1.2 V. It draws only a few microamperes, which extends the life of a single button battery for many months.

The frequency is given by the capacitor, and the LM3909 is an IC manufactured by National Semiconductor. This IC has a voltage doubler inside it that steps the battery voltage up to a level that can forward bias the LED to the point of conduction. A control circuit (relay, sensor, etc.) can be inserted between the positive pole of the battery and pin 5.

Block 249   LED Lamp Flasher

The block shown in Figure 15.4 can be used to flash two LEDs alternately or drive a lamp with currents up to 1 A (if the transistor stage is used). The frequency depends on the capacitor value and the adjustment of P1. Flashes from few seconds to one or two minutes can be produced with the values given in the circuit.

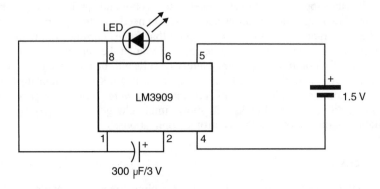

Figure 15.3   Block 248: low-voltage LED flasher.

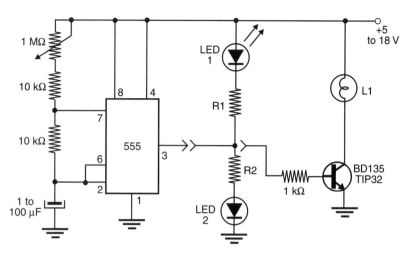

Figure 15.4   Block 249: LED/lamp flasher.

If you want to control lamps that draw currents above 200 mA, the transistor must be installed on a heat sink. If you just want to drive a single LED, either of them can be removed.

The value of R1 and R2 depends on the power supply voltage according to the following table:

Supply Voltage	R1, R2
5 V	330 Ω
6 V	470 Ω
9 V	820 Ω
12 V	1 kΩ
15 V	1.5 kΩ

Block 250   Fluorescent Lamp Inverter

Yes, fluorescent lamps can be used in robots and other projects—even devices powered from cells or batteries. With the inverter shown in Figure 15.5, the low dc voltages of cells and batteries can be converted to high enough voltages to drive a fluorescent lamp.

This circuit drains a current between 50 mA and 500 mA, depending on the lamp, the power supply voltage, and the transformer. Resistor R1 must be adjusted in the range of 1 to 10 kΩ to get the best performance.

The transistor must be mounted on a heat sink, and any transformer with a secondary winding rated between 6 and 12 V (CT) and between 100 and 500 mA can be used.

Caution: the voltage in the circuit output is very high and poses a severe shock hazard. Use insulated wires in all connections, and shield all exposed parts against accidental touch.

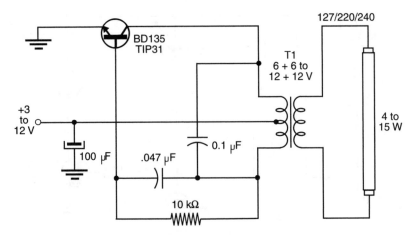

Figure 15.5   Block 250: fluorescent lamp inverter.

Block 251   Fluorescent Lamp Flasher

The block shown in Figure 15.6 uses two 555 ICs to produce high-voltage intermittent pulses to a fluorescent lamp. The transformer is the same as in the previous block. C2 must be chosen to get the best lamp brightness according to the performance of the transformer. The flash rate is adjusted by the 1 MΩ potentiometer, and the range depends on C1. The duty cycle is 50% here, but this can be altered if IC2 is controlled by other blocks via pin 4 of the 555.

Block 252   LED Sequencer

The block shown in Figure 15.7 makes ten LEDs flash in sequence at a speed determined by the adjustment of P1 and also by the value of capacitor C1. R1 depends on

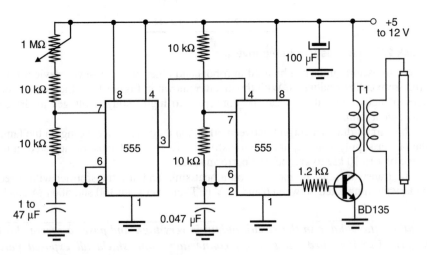

Figure 15.6   Block 251: fluorescent lamp flasher.

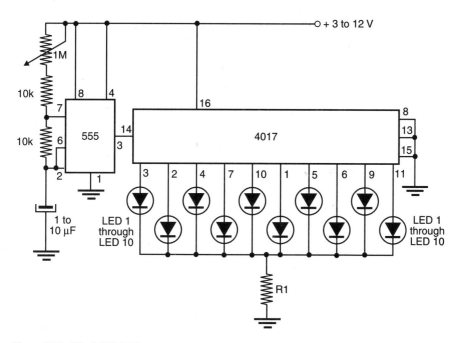

Figure 15.7   Block 252: LED sequencer.

the power supply voltage, limiting the current across each LED. The values are the same as those given in the table of Block 249.

As described in Block 219, sequences using a different number of LEDs (between 2 and 9) can be programmed using the same 4017. Special sequences of LEDs can also be programmed using Block 225.

Block 253   Driving Incandescent Lamps

Power blocks as shown in Chapter 3 can be used to drive incandescent lamps from the output of a sequencer, e.g., Block 252 and others (adapted from Block 219). Figure 15.8 shows in particular two circuits that can be used to drive lamps with currents of 500 mA (Figure 15.8a) and up to 3 A (Figure 15.8b).

The transistors must be mounted on heat sinks. Note that the lamp voltage must be the same as the voltage used to power the sequencer IC.

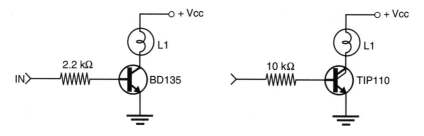

Figure 15.8   Block 253: driving incandescent lamps.

This block powers the lamp on when the input logic level is high. You can also power the lamp on at the low logic level if the blocks use an NPN transistor, as do many shown in Chapter 3.

It is also possible to drive lamps powered from the ac power line using blocks with SCRs and triacs. But remember that *the control circuit is not isolated from the ac power line.*

If you want to control high-power lamps powered from the ac line and incorporate isolation, use the circuit in Block 245. The input resistor (330 Ω) must be changed according to values given in the table of Block 249.

Block 254   Bar/Dot Display Driver

Dot/bar indicators are LED scales that can indicate voltages and other magnitudes. They can be used to indicate the action of sensors or as light effects in robots and mechatronic projects.

The block shown in Figure 15.9 is very simple. Being based on the LM3914 IC, manufactured by National Semiconductor, it needs only few external parts such as capacitors, diodes, and resistors.

This block can drive ten LEDs with the effect of a single dot or series of dots that form a bar. In the dot approach, only one LED is on, with which dot depending on the voltage applied to the input. In the bar approach, the number of LEDs that are switched on depends on the input voltage.

The trimmer potentiometer adjusts the voltage limits of the scale. It may be replaced by fixed resistors, depending on the application. This block can be powered with voltages from 3 to 12 V.

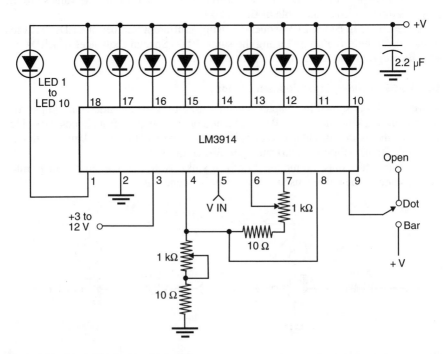

Figure 15.9   Block 254: Bar/dot display driver.

Block 255   Variable-Sound Siren

The circuit shown in Figure15.10 produces a tone that decreases in frequency from the instant the input voltage is cut. The decay time depends on the value of C1.

The circuit is triggered when a positive voltage is applied to the input. This voltage can be produced by any control block such as the ones that use TTL or CMOS logic. The value of the resistor in the input determines the opposite effect, i.e., the time period during which the sound increases in frequency.

If powered from a 3 to 6 V supply, use a BC558 or other PNP general-purpose transistor for Q2. If powered from a 9 to 12 V supply, use the BD136 or TIP32 mounted on a small heat sink.

An interesting robotic application is to use this circuit with the output of sensors or a remote control receiver. The circuit will give a audible answer to any commands. A volume control can be added in the form of a 100 Ω potentiometer in series with the loudspeaker.

Block 256   Dual-Tone Siren

The block shown in Figure 15.11 can directly drive piezoelectric transducers (high impedance), or loudspeakers with the addition of a transistor stage. The interval between the tones, and their length, are determined by C1, and the tones themselves are varied by P2. R2 can be replaced by a 100 kΩ potentiometer in series with a 10 kΩ resistor if you want a tone adjustment.

If the circuit is powered from voltage sources greater than 6 V, the transistor must be a BD135 mounted on a small heat sink. You can also place an LED at the output of the first oscillator to add a visual effect to the circuit.

When using a piezoelectric transducer, pin 1 of the 4093 can be used as an external control. The circuit is enabled by this input when it is set to the high logic level.

This circuit is an example of application of the CMOS IC 4093 in sound effects circuits. The reader can find many other 4093 configurations for sound effects in the book *CMOS Projects and Experiments* (Newnes).

Block 257   Sound and Light

The block shown in Figure 15.12 produces sound and light effects. The low-frequency oscillator formed by IC1-a drives two other oscillators, with each producing

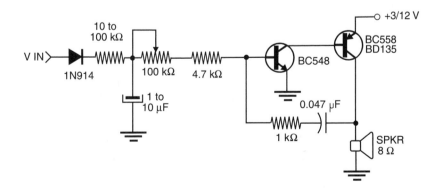

Figure 15.10   Block 255: variable-sound siren.

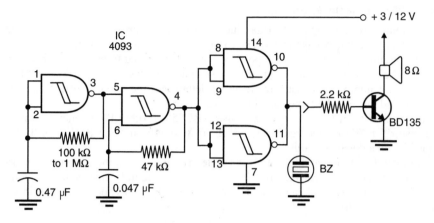

Figure 15.11   Block 256: dual-tone siren.

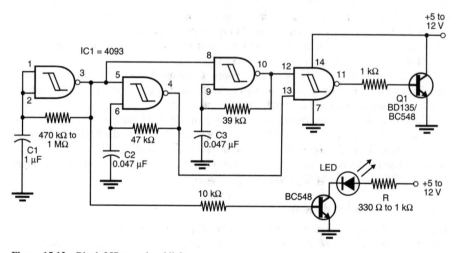

Figure 15.12   Block 257: sound and light.

a different tone. The tones are reproduced alternately by the loudspeaker. At the same time, a lamp or LED is driven, and it flashes at the same rate as the tone changes.

The flash rate and the tone changes are determined by C1. The two tones are determined by C2 and C3. The reader is free to vary these components within the range suggested in the block. If the circuit is powered with higher than 6 V, use the BD135 for the output transistor and mount it on a heat sink.

Block 258   High-Voltage Defense

High voltage discharge is used in defensive apparatus such as the ones designed for use by women in case of attack. A small inverter powered from common cells can produce discharges of 2 kV or more when activated. The shock is strong enough to repel the attack, giving the victim time to call for help.

The block shown in Figure 15.13 can be used in robotics, producing a shock if the robot is touched by an attacker. The circuit is powered from supplementary cells or the battery used to power the device. The output voltage depends on the transformer and is typically between 300 and 600 V. The reader should note that the output voltage is not the voltage rated for the transformer, since the waveform is not sinusoidal. This means that the voltage spikes can reach values much higher than the 220 or 240 V rating of the transformer. We should also point out that, despite the high voltage value, the shock is not dangerous, since the current is very low.

Of course, *due care must be taken when using this circuit.* Individuals who wear cardiac pacemakers should not touch the robot, and the designer should take precautions to limit the shock duration to a few seconds.

Block 259   Solenoid Gun

A miniature gun can be powered by a common solenoid as shown in Figure 15.14. Of course, this gun is not powerful enough to destroy anything, but the visual effect can be interesting for demonstrations, or and it can be used to build a toy battle robot that can destroy card castles and other fragile structures.

The idea is to pop out a projectile (beanbag, paper ball, wooden ball, etc.) when the core of a solenoid is engaged by magnetic forces. Using a 1 A × 12 V solenoid, you can generate enough power to project a small ball over a distance of a few

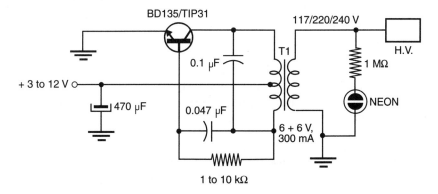

Figure 15.13   Block 258: high-voltage defense.

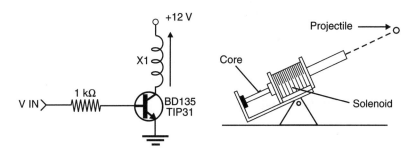

Figure 15.14   Block 259: solenoid gun.

meters. You can add a mechanism to refill the tube with a new ball each time the gun is fired.

Other possibilities include adding sound to the circuit, or even a flashlight for added realism. A variety of our blocks can be used to drive the circuit, as it needs only few milliamperes to control the transistor. Since the transistor will be on only for the fraction of second when the circuit is triggered, the heat sink is not needed.

Block 260   Laser Gun

Laser pointers can be used to simulate a gun or to give other special abilities to a robot. For example, a robot or mechatronic device can be programmed to point at objects or figures on a screen.

The circuit shown in Figure 15.15 is a simple control used to power a small laser pointer module. It uses a 3 to 12 V power supply.

The value of resistor R depends on the power supply voltage according to the following table:

Vcc (Volts)	R ($\Omega \times$ W)
3 V	10 Ω x 1 W
6 V	22 Ω x 2 W
12 V	47 Ω x 2 W

The transistor must be mounted on a heat sink. To power on the laser module, you just need to apply a positive voltage to the input of the circuit. It needs about 1 mA to power on the laser.

Block 261   General-Purpose Power Supply

A highly useful block for all who design experiments and projects involving robotics and mechatronics is a general-purpose power supply. The circuit shown by Figure 15.16 can deliver voltages in a range between 1.2 and 25 V, at currents up to 3 A.

The potentiometer adjusts the output voltage. A voltmeter is wired in parallel with the output to allow the user to measure the voltage delivered to the load. Cheap moving-coil meters can be used for this task, or even a multimeter. Digital modules such

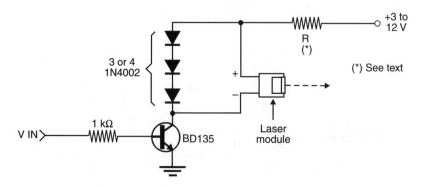

Figure 15.15   Block 260: laser gun.

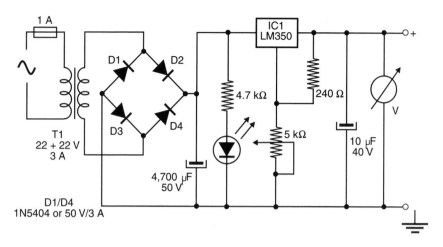

Figure 15.16   Block 261: general-purpose power supply.

as those based on the IC ICL7107 are common and cheap, giving a precise 3.5-digit indication of voltage on a liquid crystal display. The IC (LM350T) most be mounted on a heat sink, and the fuse is fundamental for protecting the circuit in case of a short.

Block 262   Battery Charger

Rechargeable batteries are a primary energy source for robots and mechatronic devices, especially if the device needs to move freely and cannot be plugged into an ac outlet. The circuit shown in Figure 15.17 is a simple battery charger for NiCad or other rechargeable batteries.

The value of resistor R is calculated according to the current needed to recharge the battery. The reader must determine the proper current and calculate R according to the following formula:

$$R = \frac{1.25}{I}$$

where   R = resistor value in ohms
         I = charging current in amperes

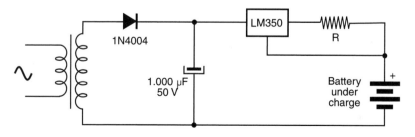

Figure 15.17   Block 262: battery charger.

For example, a 100 mA battery needs a 1.25/0.1 = 12.5 Ω resistor.

The power dissipation is calculated by the following formula:

$$P = R \times I^2$$

where   P = dissipated power in watts
          R = resistance in ohms
          I = current in amperes

For our 100 mA batteries, we have:

$$P = 12.5 \times 0.1 \times 0.1 = 0.125 \text{ W}$$

Adopt a resistor rated with twice the calculated dissipation, 0.5 W in this case.

A switch can be used to incorporate resistors in the circuit, allowing it to match batteries that the reader needs to charge. Several batteries can be placed in the charger in series for simultaneous charging as long as the sum of their voltages does not exceed 15 V.

15.4   Additional Information

Light and sound effects can be implemented using electronic circuits in a wide variety of configurations. By consulting handbooks, data sheets, and application notes from many manufacturers, we have selected some important circuits that can be used in projects that incorporate sound and light.

Because of space considerations, it is impossible to cover all of the circuits for all the applications that we have explored. Therefore, this section is devoted to suggesting components that the reader might want to use in deriving additional blocks for robotics and mechatronics applications.

LM3909

This IC, from National Semiconductor, is a 1.3 V flasher. Depending on the application, it can be used to drive LEDs from power supplies in a voltage range of 4.5 to 40 V. Figure 15.18 shows a typical application for this circuit.

The main attraction of this circuit is that, when powered from a 1.5 V cell, it operates at about 50 μA, which extends the battery life to months. More information about the uses of this IC can be found in Application Note AN-154, available from the web site of National Semiconductor (http://www.national.com).

SLB0587

The SLB0587 (from Infineon) is a touch control for ac loads. This circuit steps up and steps down the power applied to an ac load when you touch a sensor. The controlled power depends on the triac you use. Because this circuit can control inductive loads, it can also be used to control universal ac motors.

15.4.1   Musical Modules

Many toys contain small electronic modules that use recorded songs and sounds. A useful concept for robotics and mechatronics designers is to use these blocks to add

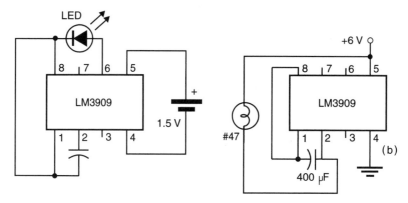

Figure 15.18   Using the LM3909 IC.

sounds to their projects. Figure 15.19 shows how to step down the voltage from a 3 to 12 V supply and power these circuits to drive a small loudspeaker.

15.4.2   Other Circuits

Other circuits that can be used to add special sound and light effects include:

- *Solid state recorders.*   With these, you can record a few words and, in some de-signs, select the ones you want to be reproduced at any particular time.
- *Smoke machines.*   When activated, these produce a cloud of inoffensive gases.
- *Stroboscopic xenon lamps.*   Many suppliers include this kind of lamp in lists of accessories for emergency situations. The circuits can be adapted to be controlled by or installed in mechatronic devices or robots.

15.5   Suggested Projects

1. Design a robot that can produce a sound each time you send it a command. Make it also produce a sequential string of LED flashes while in operation.
2. Create any mechatronic project (arm, elevator, etc.) that includes light and sound effects.

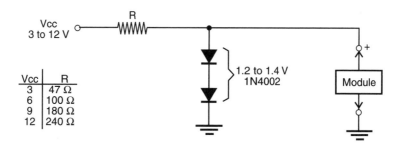

Figure 15.19   Stepping down 3 to 12 V supplies.

3. Design an artificial neuron that can learn to recognize specific sounds using a tone recognizer (Block 192), an electronic neuron employing the CA3140 or equivalent, and an oscillator block for conditioning.
4. Try to add a bar or dot display to a robot to indicate the amount of power expended when performing a task.
5. Create a robot that can be used to scare dogs, rats, mice, and other living beings who can hear ultrasonics.

15.6   Review Questions

1. What is a sequential light?
2. What is the stroboscopic effect?
3. Can we use a vox (voice-operated) circuit to control a robot?
4. How can high voltage be generated from batteries and cells?
5. How can we record voices and other sounds on a chip?
6. Are high voltages always dangerous?
7. How can NiCad and other batteries be recharged?

16

Working Safely

16.1 The Importance of Safety

You might not expect that your experiments with robots and mechatronic devices would require you to observe Asimov's "laws of robotics." But, in fact, the simplicity of these projects, the use of common parts, and the relative inexperience of the designer increase the likelihood of an accident. Yes, your simple robot, mechatronic arm, or other device can cause harm.

Table 16.1 Isaac Asimov's Laws of Robotics (1950)

0. A robot may not injure humanity or, through inaction, allow humanity to come to harm.
1. A robot may not injure a human being or, through inaction, allow a human being to come to harm, except where that would conflict with the 0th law.
2. A robot must obey the orders given it by human beings except where such orders would conflict with the first law.
3. A robot must protect its own existence as long as such protection does not conflict with the first or second law.

In simple terms, this means that you are not immune to accidents, e.g., having a finger caught in a gear, being cut by a gripper, or receiving a electric shock when touching a high-voltage circuit in one of your projects.

Although this book is not intended for the professional robot designer and therefore excludes very high-power circuits that are most likely to cause a fatal accident, it is still important to follow some basic rules of safety. The small robots and mechatronic devices described herein can move among people and contact them. If you lose control of these devices, they can do harm. When working in this field, you must be careful to avoid dangers such as a robot that collides with a child or an experimental elevator that pinches the finger of a careless operator.

Because there is a potential for accidents when working with experimental robots, and because this book provides only an introduction for readers who may want to expand on the projects, it is appropriate to include a short chapter about safety.

16.2 Experimental and Industrial Robots

Safety in the field of robotics has become more important as industrial robots have increased in importance and number. At the same time, there has been a rise in the number of accidents involving maintenance personnel and other workers. A document released by the NIOSH (National Institute for Occupational Safety and Health, www.cdc.gov/niosh/homepage.html) presents safe maintenance guidelines for robotic workstations.* Although this document is primarily directed at people who maintain industrial robot installations, the basic procedures given in its pages are valid for everyone who works with mechatronic devices and robots, including the ones you can build with the blocks described in this book.

We need to bear in mind that there are basic differences between industrial robots and the experimental mechanisms that the reader can create using common parts, many of which are home made. Robots used in industrial applications are very complex devices, and most are designed to perform a specific task such as soldering car parts in an assembly line, pick-and-place operations, workpiece orientation, etc. Taking a brief look at these machines and comparing them to small experimental robots, we find three points at which the differences are evident.

- Power level
- Degree of complexity
- Working environment

Power. Industrial robots are powerful units, and in many cases they can handle heavy pieces of metal or other materials, lifting and placing them in a specified position. These machines are very dangerous because, if they run out of control, they are powerful enough to cause severe injury or death. They can injure by impact, puncture, pinching, dragging a person over a sharp object, or pushing an operator into another machine.

A small experimental robot probably can't push you against a wall or pinch you in sensitive locations just by moving its arm, but it still can cause injury. The low power involved in experimental robot generally limits the severity of damage it can cause, but you can't ignore the possibility of losing a finger if it is caught in the drive gears of the device.

Complexity. The degree of project complexity also should be considered in a discussion about safety. As the complexity of a circuit (including the number of functions it performs) increases, there is a corresponding increase in the probability of unintended movement or an unforeseen response.

Many precautions are taken with industrial robots to avoid endangering persons who work with them, and in particular maintenance personal. Emergency switches often are installed that cut power if the unit goes out of control. In addition, sensors are sometimes installed that detect the presence of a human in a dangerous location, allowing the machinery to shut down or otherwise avoid a hazardous situation. When working with an experimental robots, the danger is not as significant, but we still must take measures to avoid accidents if the device runs out of control or changes to an unexpected operational state.

Environment. Industrial robots are very specialized machines, operating in environments that are off limits to everyone but authorized, trained personnel. These people are trained to work in dangerous places and know what to do in case of an emergency.

---

* *Safe Maintenance Guidelines for Robotic Workstations,* NIOSH Publication No. 88-108, www.cdc.gov/ niosh/88-108.html. See also *Preventing the Injury of Workers by Robots,* NIOSH Publication No. 85- 103, www.cdc.gov/niosh/85-103.html.

But, in general, experimental robots and mechatronics devices are not operated by people who have been trained to handle emergency situations. In fact, in most cases, these devices operate among average people, such as in a room with children or students or in your home or office. Although the potential for fatal hazards is lesser than with industrial robots, the casual environment makes it important to double our efforts to implement safety measures to avoid accidents.

16.3 Safety Rules

Accidents caused by a robot or a mechatronic device tend to have three main sources, as discussed below.

Mechanical. Accidents involving the mechanical parts of a robot or mechatronic device generally are caused by gears, wheels, grippers, and other moving parts that can cut, stab, or crush parts of the human body.

To avoid such accidents, some basic rules are recommended.

1. Shield any potentially hazardous moving part to avoid contact with any part of your body. Gears, wheels, chains, and other parts must be covered. If you want these parts to remain visible, you can use pieces of transparent plastic, acrylic, or other materials as shields.
2. Consider the use of sensors to detect abnormal operation of a robot that might create an accident. For example, a stalled gear or wheel can increase the current in a motor or other device. Sensing the current increase can tip us off that something is caught by the mechanism. Figure 4.25 (Chapter 4) shows how a current sensor can be added to detect when a motor is running in an overloaded condition.
3. Make sure your project is free of dangerous parts such as sharp points and edges that can stab or cut people.
4. Carefully study the environment in which your device will operate. Don't allow the mechanism to encounter objects that can be trapped by moving parts; remove all objects that can cause accidents. If possible, operate the device in a room that is free of all objects except those with which it is programmed to interact.

Electronic. The electronic circuits are not particularly dangerous, as they are placed inside the device and generally are powered from low-power (and low-voltage) sources. However, some safety measures must be considered. The following are some safety rules that are recommended to protect operators from shock and other hazards caused electrical and electronic part failures.

1. Include emergency switches for shutting down all power supplied to the circuit in case of any problem. (This measure also helps to prevent accidents involving mechanical parts.)
2. Protect all areas where high voltages are present. As possible, avoid the use of metallic parts or metallic enclosures with high-voltage circuits.
3. Include fuses or current limiting circuits in all sensitive circuits to avoid short-circuit problems.
4. Consider the use of double protection for critical circuits, especially if the robot or mechatronic device will be operated by people who don't fully understand its workings (especially children).

5. If the mechanism employs defensive devices or circuits that can cause injury to anyone, be sure that it is operated only under controlled conditions, and install redundant protective devices for shutting it down in any emergency situation.

Chemical.   The existence of chemical substances in any part of a robot or mechatronic device will create additional hazards. Substances inside a battery, gases produced by chemical cells or lead-acid batteries, and materials used to produce special effects (smoke machines, fire extinguishers, etc.) can cause accidents if used improperly.

The basic rules to prevent accidents with chemical substances are as follows:

1. Provide a secure exhaust for any gas produced inside the robot or mechatronic device.
2. Be sure that any chemical product used to produce special effects will not be injurious to humans in the case of accidental exposure or contact with the eyes or skin.
3. Have a fire extinguisher in the operating environment if any substance used in your robot is flammable.
4. Avoid using chemical-based special effects in closed or unventilated rooms.

Artificial intelligence.   Circuits and software used in artificial intelligence are dangerous in the sense that they control mechanical parts that can cause accidents. If a device runs out of control because of a bad "decision," it can cause any of the problems cited above. Such accidents can be avoided by employing the same rules plus one more: Make sure you do not design a machine that is more intelligent than you are; you can lose control.

Answers to Review Questions

Chapter 1

1. Robotics is a branch of mechatronics. Mechatronics includes all devices that operate with the combined technologies of electronics and mechanics. Robotics studies only a particular kind of device: the robot.
2. *Robot* is a word derived from a Czech novel. It means *mechanical worker* or *slave*.
3. SMA is the abbreviation for *shape memory alloy*. It is type of material that changes its shape when exposed to an electric current. It can be used as a "muscle" in projects involving electronics and mechatronics.
4. Simples eyes in robots or mechatronics devices can be made with photo sensors such as light dependent resistors (LDRs). These include CdS cells, phototransistors, and others.
5. Intelligence by software resides in a program, and intelligence by hardware is a property inherent to the electronic circuit.
6. The Turin test is made to verify if a machine is intelligent or not. It involves a dialog between human and machine.
7. Dynamic control refers to real-time control of processes, or the control of movement.
8. Your automatic garage door opener and washing machine are examples of appliances that employ mechatronics resources.
9. A neural network is a network formed by electronic neurons or cells in a configuration that can "learn."
10. A gripper is the "hand" of a robot.

Chapter 2

1. The direction of a dc motor depends on the direction of the current flowing through it.
2. The speed of a motor depends on the amount of current flowing through it.
3. Gearboxes are devices intended to reduce the speed of a motor and increase its torque.
4. A silicon diode is a semiconductor device that conducts current in only one direction.
5. The typical voltage drop in any silicon diode (e.g., the 1N4002) when forward biased is between 0.6 and 0.7 V.
6. The voltages are summed. If two 3 V batteries are wired in series, the final voltage is 6 V.

Chapter 3

1. The load is isolated from the control circuit.
2. The basic difference is in the direction of the current flowing across the two types of devices; they are opposite.
3. Voltage.
4. The input impedance of a power MOSFET is very high (hundreds or thousands of megohms).
5. To turn on an NPN bipolar transistor, we must apply a positive voltage to the base (referred to the *emitter*).
6. Rds is the resistance between drain (d) and source (s) when the transistor is on.
7. The gain of a bipolar transistor is the factor by which the collector current exceeds the base (control) current; i.e., gain = collector current/base current.

Chapter 4

1. Two transistor are conductive at the same time when an H-bridge is on.
2. At least two transistors at least are used in a half bridge.
3. The power supply is shorted out.
4. To avoid the forbidden state.
5. Four states are possible in a two-input H-bridge. Two of them determine the direction of the motor.
6. Yes, because they are compatible devices in this application.

Chapter 5

1. Half of the total power is converted into heat—in this case, 6 W.
2. A rheostat is a variable resistor, used to control the amount of power applied to a load.
3. In linear power controls, the transistor acts as a rheostat or variable resistor.
4. Yes, because when you alter the output voltage of a circuit, the amount of power applied to a load also changes.
5. It is a source that keeps the current flowing across a load constant.
6. *Duty cycle* is the ratio between the time a circuit is on and the duration of the complete cycle, expressed as a percentage.
7. The power in the load is zero.
8. Because, with 100% power, the time off must be zero, and this will stop the oscillator.

Chapter 6

1. An SCR is a half-wave device, since it conducts current only in one direction.
2. We must apply a positive voltage to the gate.
3. A Quadrac is a device formed by a triac and a diac in the same package.
4. The holding current is the minimum current accross a SCR that keeps it in the conductive state.
5. The snap action of an SCR during turn-on and turn-off generates high-frequency signals.
6. It is called a *crowbar*.
7. The suffix D means that the device is rated for 400 V.

Chapter 7

1. SMA indicates *shape memory allow,* a material that changes its form when exposed to electric current.
2. The main material used to make SMAs is *flexinol.*
3. The length increases with temperature.
4. The force of a solenoid is determined by the amount of current flowing through it and the number of turns of wire in its coil.
5. The magnetic field is more intense inside the solenoid.
6. We divide the voltage applied to the solenoid by the coil's resistance.

Chapter 8

1. The difference is in the way the shaft moves: a stepper motor moves in steps, and the shaft can be put in any position by an external control.
2. In some cases, yes.
3. Yes.
4. The two-phase stepper motor uses two coils.
5. A *translator* is a circuit that generates the necessary current to drive a stepper motor from external control signals.
6. By maintaining the current across a coil, the shaft is locked in position.
7. The direction of a stepper motor can be reversed by changing the sequence of pulses applied to its coils.

Chapter 9

1. *Debouncing circuits* are needed to avoid the effects of voltage spikes that are generated when a switch or sensor is turned on and off.
2. To increase their operating life (because there is no oxygen to burn the contacts).
3. Yes, because the load current and voltage are below the specifications for this device.
4. Constant-duration pulses can be generated from switches and sensors using monostables.
5. This means that the switch can control one circuit, turning it on and off (single-pole, single-throw).
6. The normally open switch can be used to turn off a load if it is wired to control a relay with normally closed (NC) contacts.
7. The maximum current drained or delivered by a CMOS output powered from a 10 V supply is 2.25 mA.

Chapter 10

1. A transducer is a device that converts one form of energy into another. A loud-speaker is a transducer that converts electrical energy into sound (acoustic energy)
2. The electrical resistance is reduced.
3. Using devices as NTCs or PTCs.
4. Because they can react to dissipate and absorb heat from the environment faster than the larger devices.
5. A robot can see in the dark if we flood its operating environment with infrared light.

Chapter 11

1. An operational amplifier has two inputs: inverting (–) and a noninverting (+).
2. The phase of the signal is shifted 180° (inverted).
3. The output goes to a high logic level or to the positive voltage of the power supply.
4. Two comparators are used to perform as a window comparator
5. In the package of an LM324, we find four comparators.
6. It is a power supply that delivers a positive and a negative voltage, referenced to ground.

Chapter 12

1. A remote control can use wires or any other medium to send signals to the controlled unit. A *wireless* control uses no wires (e.g., radio or light) to send signals to the controlled unit.
2. The mobility of the controlled unit is limited by the length of its wires.
3. Noise in the ambient can cause interferences.
4. Electromagnetic interference (EMI) can be picked up by the receivers used in remote control systems that use radio waves.
5. A PLL is a *phase locked loop*, a circuit that can be used to select signals or a specific frequency.
6. Hybrid modules are circuits that use discrete devices mounted on a silicon chip to perform specific functions, e.g., receivers, transmitters, etc.
7. Because we need less power, and the circuits are less sensitive to EMI.

Chapter 13

1. The monostable circuit remains in the triggered state only for a predetermined time interval. In the bistable circuit, the state changes only when a trigger pulse is received.
2. Logic circuits can be wired together to make complex logic decisions.
3. The counter can be used to trigger a sequence of circuits that are wired to perform logic decisions.
4. The 555 as a monostable can be wired in a cascade configuration to perform multiple timer tasks.
5. The 4013 is a dual CMOS D-type flip-flop.
6. A DAC, or *digital-to-analog converter,* is a circuit that converts digital information into analog signals such as voltages and currents. It can be used to control dc circuits from digital logic.
7. A VCO is a *voltage controlled oscillator,* a circuit that produces a signal with a frequency that depends on the voltage applied to the input.

Chapter 14

1. *Intelligence* can be defined as the ability of a system of learn, make logic decisions, and generate feelings.
2. Intelligence by software resides in a program, and intelligence by hardware is a property inherent to a circuit.
3. A circuit can learn by modifying some of its characteristics according to experience (use).

4. A thermal memory cell uses heat to store information about the experiences of the circuit or to learn.
5. An electronic neuron can learn by changing its response to external stimuli.
6. A microprocessor can run artificial intelligence software, thereby adding intelligence to a robot.
7. A neural network is formed by artificial neurons wired in manner that allows the circuit to learn.
8. An *analog-to-digital converter* is a circuit that converts analog signals (voltages and currents) into digital information. It can be use to interface sensors with microprocessors or other digital circuits.
9. The parallel port of a computer has eight input/output (I/O) channels.

Chapter 15

1. A sequential light is a system in which lamps flash one after the other in sequence.
2. The stroboscopic effect is the apparent freezing of a moving object when illuminated by a rapidly flashing light.
3. A vox circuit can be used to control devices through the use of sounds picked up by a microphone.
4. Inverters are circuits used to step up the low dc voltage produced by batteries and other dc sources.
5. To be recorded in memory chips, sound must be converted into digital format.
6. No, if the current is limited to very low values (a few microamperes).
7. These batteries are recharged by forcing a reverse current to flow through them (i.e., the opposite of the direction of flow when they are delivering current to a load).